THE SMART GUIDE TO

Chemistry

BY BRIAN NORDSTROM

SECOND EDITION

The Smart Guide To Chemistry - Second Edition

Published by

Smart Guide Publications, Inc.
2517 Deer Chase Drive
Norman, OK 73071
www.smartguidepublications.com

For information, address: Smart Guide Publications, Inc. 2517 Deer Creek Drive, Norman, OK 73071

SMART GUIDE and Design are registered trademarks licensed to Smart Guide Publications, Inc.

International Standard Book Number: 978-1-937636-62-3

Library of Congress Catalog Card Number:
11 12 13 14 15   10 9 8 7 6 5 4 3 2 1

Printed in the United States of America

Cover design: Lorna Llewellyn
Copy Editor: Ruth Strother
Back cover design: Joel Friedlander, Eric Gelb, Deon Seifert
Back cover copy: Eric Gelb, Deon Seifert
Illustrations: James Balkovek
Production: Zoë Lonergan
Indexer: Cory Emberson
V.P./Business Manager: Cathy Barker

ACKNOWLEDGMENTS

I would like to thank the outstanding teachers who have so greatly enriched my life over the years. Special gratitude goes to the following people: My high school physics teacher, Sidney Quigley, Berkeley chemistry professors Samuel Markowitz, Kenneth Street, Neil Bartlett, Harold S. Johnston, and Nobel laureate Glenn T. Seaborg, Berkeley physics professors George Trilling, Robert Brown, William Chinowsky, and Nobel laureate Emilio Segrè, Berkeley history professor Rafael Sealey and Berkeley wildlife biologist A. Starker Leopold.

I would also like to thank my many professional colleagues and mentors over the years, with special thanks to chemist Charles Koelsche, who always believed in me.

Lastly, but not least, I would like to thank the Smart Guide series publisher, without whose long hours of work this book would not have happened.

TABLE OF CONTENTS

PART ONE

The Nature of Matter

Periodic Table of the Elements

Z	Symbol	Name	Configuration	Atomic mass	Group
1	H	hydrogen	$1s^1$	1.008	1A
2	He	helium	$1s^2$	4.003	8A
3	Li	lithium	$[He]2s^1$	6.941	1A
4	Be	beryllium	$[He]2s^2$	9.012	2A
5	B	boron	$[He]2s^2 2p^1$	10.81	3A
6	C	carbon	$[He]2s^2 2p^2$	12.01	4A
7	N	nitrogen	$[He]2s^2 2p^3$	14.01	5A
8	O	oxygen	$[He]2s^2 2p^4$	16.00	6A
9	F	fluorine	$[He]2s^2 2p^5$	19.00	7A
10	Ne	neon	$[He]2s^2 2p^6$	20.18	8A
11	Na	sodium	$[Ne]3s^1$	22.99	1A
12	Mg	magnesium	$[Ne]3s^2$	24.31	2A
13	Al	aluminum	$[Ne]3s^2 3p^1$	26.98	3A
14	Si	silicon	$[Ne]3s^2 3p^2$	28.09	4A
15	P	phosphorus	$[Ne]3s^2 3p^3$	30.97	5A
16	S	sulfur	$[Ne]3s^2 3p^4$	32.07	6A
17	Cl	chlorine	$[Ne]3s^2 3p^5$	35.45	7A
18	Ar	argon	$[Ne]3s^2 3p^6$	39.95	8A
19	K	potassium	$[Ar]4s^1$	39.10	1A
20	Ca	calcium	$[Ar]4s^2$	40.08	2A
21	Sc	scandium	$[Ar]4s^2 3d^1$	44.96	3B
22	Ti	titanium	$[Ar]4s^2 3d^2$	47.88	4B
23	V	vanadium	$[Ar]4s^2 3d^3$	50.94	5B
24	Cr	chromium	$[Ar]4s^1 3d^5$	52.00	6B
25	Mn	manganese	$[Ar]4s^2 3d^5$	54.94	7B
26	Fe	iron	$[Ar]4s^2 3d^6$	55.85	8B
27	Co	cobalt	$[Ar]4s^2 3d^7$	58.93	8B
28	Ni	nickel	$[Ar]4s^2 3d^8$	58.69	8B
29	Cu	copper	$[Ar]4s^1 3d^{10}$	63.55	1B
30	Zn	zinc	$[Ar]4s^2 3d^{10}$	65.39	2B
31	Ga	gallium	$[Ar]4s^2 3d^{10} 4p^1$	69.72	3A
32	Ge	germanium	$[Ar]4s^2 3d^{10} 4p^2$	72.64	4A
33	As	arsenic	$[Ar]4s^2 3d^{10} 4p^3$	74.92	5A
34	Se	selenium	$[Ar]4s^2 3d^{10} 4p^4$	78.96	6A
35	Br	bromine	$[Ar]4s^2 3d^{10} 4p^5$	79.90	7A
36	Kr	krypton	$[Ar]4s^2 3d^{10} 4p^6$	83.79	8A
37	Rb	rubidium	$[Kr]5s^1$	85.47	1A
38	Sr	strontium	$[Kr]5s^2$	87.62	2A
39	Y	yttrium	$[Kr]5s^2 4d^1$	88.91	3B
40	Zr	zirconium	$[Kr]5s^2 4d^2$	91.22	4B
41	Nb	niobium	$[Kr]5s^1 4d^4$	92.91	5B
42	Mo	molybdenum	$[Kr]5s^1 4d^5$	95.94	6B
43	Tc	technetium	$[Kr]5s^2 4d^5$	(98)	7B
44	Ru	ruthenium	$[Kr]5s^1 4d^7$	101.1	8B
45	Rh	rhodium	$[Kr]5s^1 4d^8$	102.9	8B
46	Pd	palladium	$[Kr]4d^{10}$	106.4	8B
47	Ag	silver	$[Kr]5s^1 4d^{10}$	107.9	1B
48	Cd	cadmium	$[Kr]5s^2 4d^{10}$	112.4	2B
49	In	indium	$[Kr]5s^2 4d^{10} 5p^1$	114.8	3A
50	Sn	tin	$[Kr]5s^2 4d^{10} 5p^2$	118.7	4A
51	Sb	antimony	$[Kr]5s^2 4d^{10} 5p^3$	121.8	5A
52	Te	tellurium	$[Kr]5s^2 4d^{10} 5p^4$	127.6	6A
53	I	iodine	$[Kr]5s^2 4d^{10} 5p^5$	126.9	7A
54	Xe	xenon	$[Kr]5s^2 4d^{10} 5p^6$	131.3	8A
55	Cs	cesium	$[Xe]6s^1$	132.9	1A
56	Ba	barium	$[Xe]6s^2$	137.3	2A
57	La	lanthanum	$[Xe]6s^2 5d^1$	138.9	*
58	Ce	cerium	$[Xe]6s^2 4f^1 5d^1$	140.1	*
59	Pr	praseodymium	$[Xe]6s^2 4f^3$	140.9	*
60	Nd	neodymium	$[Xe]6s^2 4f^4$	144.2	*
61	Pm	promethium	$[Xe]6s^2 4f^5$	(145)	*
62	Sm	samarium	$[Xe]6s^2 4f^6$	150.4	*
63	Eu	europium	$[Xe]6s^2 4f^7$	152.0	*
64	Gd	gadolinium	$[Xe]6s^2 4f^7 5d^1$	157.2	*
65	Tb	terbium	$[Xe]6s^2 4f^9$	158.9	*
66	Dy	dysprosium	$[Xe]6s^2 4f^{10}$	162.5	*
67	Ho	holmium	$[Xe]6s^2 4f^{11}$	164.9	*
68	Er	erbium	$[Xe]6s^2 4f^{12}$	167.3	*
69	Tm	thulium	$[Xe]6s^2 4f^{13}$	168.9	*
70	Yb	ytterbium	$[Xe]6s^2 4f^{14}$	173.0	*
71	Lu	lutetium	$[Xe]6s^2 4f^{14} 5d^1$	175.0	*
72	Hf	hafnium	$[Xe]6s^2 4f^{14} 5d^2$	178.5	4B
73	Ta	tantalum	$[Xe]6s^2 4f^{14} 5d^3$	180.9	5B
74	W	tungsten	$[Xe]6s^2 4f^{14} 5d^4$	183.9	6B
75	Re	rhenium	$[Xe]6s^2 4f^{14} 5d^5$	186.2	7B
76	Os	osmium	$[Xe]6s^2 4f^{14} 5d^6$	190.2	8B
77	Ir	iridium	$[Xe]6s^2 4f^{14} 5d^7$	192.2	8B
78	Pt	platinum	$[Xe]6s^1 4f^{14} 5d^9$	195.1	8B
79	Au	gold	$[Xe]6s^1 4f^{14} 5d^{10}$	197.0	1B
80	Hg	mercury	$[Xe]6s^2 4f^{14} 5d^{10}$	200.5	2B
81	Tl	thallium	$[Xe]6s^2 4f^{14} 5d^{10} 6p^1$	204.4	3A
82	Pb	lead	$[Xe]6s^2 4f^{14} 5d^{10} 6p^2$	207.2	4A
83	Bi	bismuth	$[Xe]6s^2 4f^{14} 5d^{10} 6p^3$	209.0	5A
84	Po	polonium	$[Xe]6s^2 4f^{14} 5d^{10} 6p^4$	(209)	6A
85	At	astatine	$[Xe]6s^2 4f^{14} 5d^{10} 6p^5$	(210)	7A
86	Rn	radon	$[Xe]6s^2 4f^{14} 5d^{10} 6p^6$	(222)	8A
87	Fr	francium	$[Rn]7s^1$	(223)	1A
88	Ra	radium	$[Rn]7s^2$	(226)	2A
89	Ac	actinium	$[Rn]7s^2 6d^1$	(227)	**
90	Th	thorium	$[Rn]7s^2 6d^2$	232.0	**
91	Pa	protactinium	$[Rn]7s^2 5f^2 6d^1$	231	**
92	U	uranium	$[Rn]7s^2 5f^3 6d^1$	238	**
93	Np	neptunium	$[Rn]7s^2 5f^4 6d^1$	237	**
94	Pu	plutonium	$[Rn]7s^2 5f^6$	(244)	**
95	Am	americium	$[Rn]7s^2 5f^7$	(243)	**
96	Cm	curium	$[Rn]7s^2 5f^7 6d^1$	(247)	**
97	Bk	berkelium	$[Rn]7s^2 5f^9$	(247)	**
98	Cf	californium	$[Rn]7s^2 5f^{10}$	(251)	**
99	Es	einsteinium	$[Rn]7s^2 5f^{11}$	(252)	**
100	Fm	fermium	$[Rn]7s^2 5f^{12}$	(257)	**
101	Md	mendelevium	$[Rn]7s^2 5f^{13}$	(258)	**
102	No	nobelium	$[Rn]7s^2 5f^{14}$	(259)	**
103	Lr	lawrencium	$[Rn]7s^2 5f^{14} 6d^1$	(262)	**
104	Rf	rutherfordium	$[Rn]7s^2 5f^{14} 6d^2$	(261)	4B
105	Db	dubnium	$[Rn]7s^2 5f^{14} 6d^3$	(262)	5B
106	Sg	seaborgium	$[Rn]7s^2 5f^{14} 6d^4$	(266)	6B
107	Bh	bohrium	$[Rn]7s^2 5f^{14} 6d^5$	(264)	7B
108	Hs	hassium	$[Rn]7s^2 5f^{14} 6d^6$	(277)	8B
109	Mt	meitnerium	$[Rn]7s^2 5f^{14} 6d^7$	(268)	8B
110	Ds	darmstadtium	$[Rn]7s^2 5f^{14} 6d^8$	(271)	8B
111	Rg	roentgenium		(272)	1B
112	Cn	copernicium		(277)	2B
113	Uut			(?)	3A
114	Uuq			(285)	4A
115	Uup			(?)	5A
116	Uuh			(289)	6A
117	Uus			(?)	7A
118	Uuo			(?)	8A

Lanthanide Series*
Actinide Series**

Los Alamos National Laboratory Chemistry Division

© Copyright 2011 Los Alamos National Security, LLC. All rights reserved. The public may copy and use this information without charge, provided that this Notice and any statement of authorship are reproduced on all copies. Neither the Government nor LANS makes any warranty, express or implied, or assumes any liability or responsibility for the use of this information.

Los Alamos
NATIONAL LABORATORY

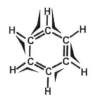

CHAPTER 1

The Wonderful World of Atoms and Molecules

In This Chapter

➤ Chemistry defined

➤ Types of chemists and what they do

➤ What it takes to be a chemist

In this chapter you will be introduced to the science and practice of chemistry. Chemistry is all about atoms and molecules, but chemistry is practiced by real people working in government, industry, and education. Although this book focuses on the science of chemistry, I'll also try to show you the human side of it.

What Chemistry Is

Chemistry is the science of atoms and molecules—the "stuff" of which everything on Earth and in the universe is made. At the microscopic level—a scale of matter too small to be visible to the naked eye—chemists study the structures of atoms, the shapes of molecules, and the forces that hold atoms and molecules together. At the macroscopic level—a scale of matter that is visible to the naked eye—chemists study the chemical and physical properties of the states of matter—solid, liquid, gas, or mixture.

Chem Vocab

Chemistry is a science concerned with matter: identifying substances, describing them, and studying their transformations.

But chemistry is not only about matter that is just sitting there. Just as importantly, if not more so, chemistry is also the science of changes, or transformations, of matter. At a nuclear level—a scale smaller than an atom itself—these changes may be the transmutation of one element into another element. At the level of whole atoms and molecules, these changes involve the processes of breaking chemical bonds and forming new bonds with the result that one set of chemical substances—or mixtures of substances—are transformed into a new set of substances that may have properties that differ completely from those of the original materials.

Finally, chemistry is not limited only to substances that exist naturally on Earth. Literally millions of new substances have been synthesized in chemical laboratories that may mimic substances found in nature, although many times these substances may have completely novel chemical and physical properties not found in nature.

Chemists are not finished making new substances. Thousands of new materials are developed every year in laboratories in countries all over the globe. Some of these new substances may lead to cures for debilitative diseases. Some substances might lead to more environmentally sustainable agricultural chemicals or to lightweight, durable structural materials. Others may result in revolutionary new forms of energy. Unfortunately, some may also become agents of war and terrorism—agents of biological and chemical warfare, novel weapon delivery systems, or explosives that can bypass even the most sophisticated high-technology security systems. Scientists never know what applications may be found for their discoveries, or what uses—for good or for evil—others will find for their discoveries.

Important Point

Chemistry is often called the central science because so many other areas of science and technology rely on chemistry as a foundation.

Chemistry: The Central Science

Chemists often like to call chemistry the central science because chemistry is fundamental to so many other endeavors—biology, medicine, agriculture, physics, astronomy, planetary science, cosmology, geology, and engineering, just to name a few. To say that chemistry is the science of atoms and molecules is really to say that chemistry is the science of everything, since all matter is composed of atoms and molecules. Examples span all sorts of substances from raindrops, dew, and mist to puddles, ponds, lakes, and oceans; from sand and gravel to rocks, boulders, and mountains; from cotton, wool, and silk to nylon, rayon, and polyester; from natural wood products to plastics, concrete, and steel; from all matter on Earth to all matter on the Moon and other planets, the Sun and other stars, and even to the space between stars and galaxies.

It is difficult to think of anything that is not chemistry. The standard exceptions are usually at the extremes of the very, very small or the very, very large. On the smallest scales of matter, chemistry usually stops at the level of the protons, neutrons, and electrons that make up atoms. It is left to elementary particle physicists to investigate the properties of matter at smaller scales than that—the quarks of which protons and neutrons are made, or other subatomic particles with exotic names like muons, pions, or mesons. While chemists investigate the nuclear reactions that take place in stars like our Sun and the chemical composition of stars and the space between stars, it is astrophysicists and cosmologists who are concerned about the large scale structures of the universe—galaxies and clusters of galaxies and how they come into being and evolve over time.

Chemistry: The Basis of a Global Industry

The millions of chemical substances—natural or artificial—that have ever been cataloged have had profound effects on the human race. Cave men and women did not know it, but their discovery of how to cook foods, especially meat, altered forever the development of modern humans. Cooking is a chemical reaction that makes food easier to chew and swallow, more digestible, and freer of harmful pathogens, all of which contributed to improved human health, increased life spans, and accelerated brain development.

Substituting metal implements for stone was chemistry. Cave men and women lived in the Stone Age, when rocks were either used as they were found or fashioned or sharpened into weapons, tools, and utensils. The Bronze Age replaced the Stone Age when more modern humans learned to work with copper. Without knowing it, these people were early metallurgists, and metallurgy is chemistry. Soon, the Bronze Age was superseded by the Iron Age, as iron replaced copper in many aspects of life.

In the eighteenth century, the industrial revolution could not have taken place without chemistry. The industrial revolution was built on the steam engine, and steam engines required energy—energy obtained from fuels through chemical processes known as combustion reactions.

Today, the chemical industry is a global industry with significant investments in many diverse fields:

➤ Fossil fuels, from coal and petroleum to natural gas

➤ Alternative energy technologies, from nuclear and solar power to ethanol and hydrogen

Important Point

On a global level, chemistry is one of the world's largest industries and includes such diverse enterprises as petroleum, pharmaceuticals, agricultural products, and basic chemicals.

➤ Pharmaceuticals, everything from analgesics as ordinary as aspirin to blockbuster drugs that fight heart disease and cancer

➤ Cosmetics, from lipstick to perfume

➤ Agricultural products, from fertilizers and pesticides to genetically modified cereal crops

➤ Synthetic materials, from plastics and fibers to rubber

➤ Basic chemicals, everything from household cleaners and wastewater treatment to the pages of this book

So what is chemistry? Everything is chemistry!

What Chemists Do

Chemistry is what chemists do. The field of chemistry encompasses a vast range of activities. Because there is so much demand for the skills of chemists in industry, government, and education, the American Chemical Society is the largest professional scientific organization devoted to a single science, with sister organizations in other countries. Chemists literally are everywhere, and many of them work in laboratories. Chemists may be researchers, teachers, consultants, sales representatives, or technicians. Chemists may hold managerial or administrative positions.

Historical Note

Alchemy was one of humankind's earliest scientific endeavors. Its origins predate any historical records.

Chemists do all kinds of interesting work. They may be analyzing moon rocks and meteorites, developing lifesaving antibiotics, synthesizing new elements, studying the chemistry of the human body, or probing into theories about the origin of life. Let's take a look at some of the subdisciplines of chemistry in which chemists engage.

Analytical Chemists

From the earliest years of alchemy, there has been a demand for people who can answer two questions:

➤ What is the composition of a substance?

➤ How much of one substance is present in a mixture of substances?

Historical Note

Alchemy was the predecessor to the modern science of chemistry. Alchemists discovered new elements and compounds and attempted to change lead into gold. Although they failed to achieve the latter, they made many valuable contributions to the early knowledge of chemical substances.

The first question belongs to the domain of qualitative analysis, which examines materials to find out what elements or compounds are in them. The second question belongs to the domain of quantitative analysis, which measures how much of each element or compound is present in a mixture of substances.

Chem Vocab

Qualitative analysis is the determination of *what* is in a substance.

Quantitative analysis is the determination of *how much* of a substance is present in a mixture.

Qualitative analysis requires that mixtures be separated into their components, which can be compounds or simple elements. Compounds, in turn, may be further separated into their constituent elements. In geology and the mining industry, for example, rocks, minerals, or soils are analyzed to find out what metals (such as copper, nickel, or titanium) or other elements (such as chlorine or phosphorus) are present. Municipal water facilities have to identify and remove any contaminants (such as arsenic, lead, or nitrates) that might be present in surface or groundwater supplies before the water enters a city's water supply system. The food industry analyzes its products so that the labels can inform consumers of what kinds of fats, proteins, carbohydrates, vitamins, minerals, and fiber are present in canned, packaged, and prepared foods. The pharmaceutical industry analyzes samples of all of its products in the attempt to ensure against contamination.

Historically, quantitative analysis involved "wet" chemical techniques—titration, filtration, and gravimetric analysis, for example. Today, most quantitative analyses are done using "instrumental" techniques. An example you might be familiar with is the smog check many municipalities require motor vehicles to pass before they can be registered. In a smog check, a hose connects the tailpipe of a vehicle to a machine that can separate the exhaust gases into different categories—carbon monoxide (CO), oxides of nitrogen (NO_x), and unburned hydrocarbons (HC)—and measure the concentration of each gas in parts per million (ppm).

Chem Vocab

Ppm is an acronymn for parts per million. For example, a 1 ppm salt solution means that there is 1 gram of salt per 1 million grams of total solution.

It makes sense that quantitative measurements follow qualitative analyses. In the example of rocks and minerals, once the metal content of ores has been established, then samples of ores can be assayed to find out what percentage of the ore is a metal of value. In the food industry example, labels on food products tell not just what is present in the foods, but how many grams or milligrams of fat, protein, carbohydrates, vitamins, minerals, and fiber are present per serving.

Organic Chemists

Organic chemistry is all about carbon and its compounds. The term *organic chemistry* originated during the nineteenth century, when chemists thought that carbon-containing substances could only be made by or in living organisms. That definition is no longer in use. Today we simply define an organic compound as any chemical compound that contains at least carbon and hydrogen. The reason for including hydrogen is to rule out carbon monoxide (CO) and carbon dioxide (CO_2), both of which contain carbon but are usually classified as inorganic substances. In addition to carbon and hydrogen, organic compounds may also contain oxygen, nitrogen, phosphorus, sulfur, and other elements.

Much of the modern chemical industry is based on organic chemistry. Petroleum, natural gas, lubricating oils, alcohols, ethers, solvents, weak acids, vitamins and other pharmaceutical products, and many synthetic polymers are all organic compounds.

Biochemists

Biochemistry is that branch of organic chemistry that deals with the chemistry of living organisms. Human bodies, as well as all plants and animals, contain large organic molecules essential to the many biological processes that define replication of organisms, growth, and metabolic activities. Important classes of biochemical substances include carbohydrates, lipids, proteins, and nucleic acids (such as DNA). Biochemists have contributed to chemical knowledge by elucidating the pathways by which photosynthesis takes place in green plants and by which respiration takes place in all living organisms. The syntheses of biologically important substances such as vitamins, cortisone, and hormones are often done by biochemists.

Biochemists are often pharmaceutical chemists. They work for companies that do the research necessary to develop and bring new drugs to the marketplace. Some of the major advances that have been made in the battle against disease have been accomplished by biochemists.

Important Point

Biochemistry, a blend of biology and chemistry, is a good example of the interdisciplinary nature of chemistry.

Inorganic Chemists

Simply stated, inorganic chemistry deals with the 117 elements in the periodic table other than carbon. The elements in the periodic table are broadly grouped into three classifications: metals, nonmetals, and metalloids (or semimetals). Inorganic chemists describe the physical and chemical properties of the elements themselves, as well as all of the chemical compounds the elements can form, both in nature and in the laboratory.

Metals—especially iron in the form of steel—are the workhorses of much of modern heavy industry. Thus, inorganic chemists work to develop methods for isolating metals from their ores and for finding new and economically useful applications for metals.

Many nonmetals and their compounds find essential uses in modern society. Life could not exist without liquid water. Most living organisms could not exist without oxygen gas. Chlorine disinfects drinking water supplies. Salt flavors our food. Neon is used in outdoor lighting. Helium gives balloons and blimps their buoyancy, although helium's cryogenic (low temperature) uses are much more important. Liquid nitrogen is also a cryogenic substance. In its liquid state carbon dioxide is an environmentally friendly solvent. Fertilizers are mostly used to supply nitrogen, phosphorus, potassium, and sulfur to plants in the form of compounds that plants can assimilate. Hydrogen is being investigated as a clean-burning, renewable fuel that could substitute for fossil fuels.

Historical Note

It is virtually impossible to imagine what modern society would be like if no one had ever discovered the semiconductor properties of silicon and germanium.

The metalloids silicon and germanium are the building blocks of computer chips and transistors. Many other elements are used in modern electronics and other technologies. Artificial radioactive substances are produced and used in hospitals every day both for medical diagnosis and for medical treatment. In short, unless an element is particularly rare or expensive, just about every element in the periodic table finds some use in our society.

Physical Chemists

Physical chemists apply the methods of physics to an understanding of the variables that affect how chemical reactions occur so that chemists can control the rates at which chemical reactions take place and improve the yield of products. Three of the major subfields of physical chemistry are quantum chemistry, thermodynamics, and chemical kinetics.

In quantum chemical studies, chemists relate the structures of atoms and molecules at the most fundamental microscopic level to the observable properties of elements and compounds at the macroscopic level. Quantum chemists study the nature of chemical bonding, the forces that hold molecules together in the solid and liquid states, and how the shapes of molecules determine the properties of those substances.

Thermodynamics is the study of chemical equilibrium and of the heat effects associated with chemical reactions. Very importantly, thermodynamicists are employed to maximize the yields of synthetic industrial and pharmaceutical chemicals.

Historical Note

Before the beginnings of modern physics in the early twentieth century, the three fundamental subdisciplines of the physical sciences were classical mechanics, electromagnetism, and thermodynamics. Much of the accomplishments of the industrial revolution were derived from these fields of science.

Chemical kinetics is the study of the factors that control the rates at which chemical reactions take place. Kineticists investigate the effects of changing pressure, concentration of reactants, and temperature on the rates of chemical reactions. Many reactions take place too slowly for industrial applications at ambient temperatures and pressures, so kineticists find ways to speed up economically useful chemical reactions.

One way to speed up chemical reactions is to employ catalysts, chemical substances that speed up chemical reactions but that remain unchanged by the reactions. The best catalysts in the world are biochemical catalysts—the enzymes found in all living organisms. For many industrial processes, kineticists seek to find catalysts that at least approach the efficiency of naturally occurring enzymes.

Nuclear Chemists

Another branch of physical chemistry is nuclear chemistry. Nuclear chemists work with radioactive materials, which may occur naturally or be produced artificially in nuclear reactors. Nuclear chemists study the properties of these substances and investigate ways in which radioactive materials may be useful in a wide range of applications, including medicine and agriculture among other fields.

For the past three-quarters of a century, nuclear chemists have synthesized a total of 28 elements that do not occur in nature but occupy positions in the periodic table after the element uranium. As of 2012, a total of 118 elements are known. Nuclear chemists continue to work at extending the periodic table to even heavier elements.

Important Point

Almost one-quarter of the chemical elements do not occur on Earth naturally. They have been synthesized in laboratories.

Chemical Engineers

Chemical engineers are applied physical chemists. Chemical engineers usually work in industry. Their job is to take benchtop synthetic methods that are usually done on the scale of a few grams of a substance and scale the processes up to producing metric tons of the same substances. The ability to apply the principles of thermodynamics and chemical kinetics to real-life problems of manufacturing useful chemical substances on a large scale is the trademark of a chemical engineer.

Educational Requirements to be a Chemist

Chemists generally have a college degree. Community colleges may offer two-year certification programs to be a laboratory technician. Most four-year colleges and universities offer a bachelor's degree in chemistry, and some universities may also offer a bachelor's degree in chemical engineering. Colleges and universities with graduate programs may offer master's and doctor's degrees in chemistry. Entry-level positions in research or sales usually require a bachelor's or master's degree. To be a leader of a research program usually requires a Ph.D. in chemistry or chemical engineering.

Teaching chemistry at the high school level usually requires a bachelor's degree in chemistry, or a degree in a related science such as biology or physics with a strong chemistry background. Teaching chemistry at a community college requires a master's or doctor's degree in chemistry. Teaching at a four-year college or university almost always requires a Ph.D., especially if there is an expectation that faculty conduct research programs.

Regardless of an undergraduate's specific plans for a career in chemistry or other related field, the sequence of required college courses is fairly standard:

➤ One year each of general, organic, and physical chemistry

➤ One semester each of analytical chemistry, advanced inorganic chemistry, and electives that might include physical organic chemistry, advanced organic, and inorganic chemistry

➤ Biochemistry, or nuclear chemistry

➤ One to two years of mathematics, including calculus and differential equations

➤ One year of physics

➤ General education courses that include English composition, speech, computer science, the social sciences, and the humanities

➤ Electives that might include additional mathematics or physics, plus biology, geology, or other areas of interest

Laboratory work is a major component of an undergraduate education in chemistry. The courses taken during the first three years usually include laboratory work. In the senior year a student may take advanced laboratory courses or participate in research under the supervision of a faculty member. An undergraduate research experience is especially important for students intending to continue on to graduate programs. It is not unusual for undergraduate students to have already published a paper or two in a refereed research journal before graduating.

Here is some advice to remember if you are a student. You never know where your career path may lead you. A required course today may be the ticket to success on the job at some future date. Get the most you can out of every class you take. Get to know your professors. Believe it or not, they are real people, and you never know when you might want a letter of recommendation from one of them someday! And don't neglect those general education courses. No matter what you do in life, the ability to communicate—and to communicate well—is essential. Master both oral and written communication skills, and you will find more doors opening to you down the road.

Conclusion: Chemistry as a Human Enterprise

Chemistry does not get done all by itself. People do chemistry. Chemistry, together with all of the other natural sciences, is as much a human enterprise as are the humanities or the social sciences. The next chapter is intended to present chemistry within its historical context, highlighting some of the major figures who have contributed to chemistry over the centuries.

CHAPTER 2

A Brief History of Chemistry

In This Chapter

➤ Early chemistry

➤ Alchemy

➤ Modern chemistry

➤ Chemistry in the twentieth and twenty-first centuries

In this chapter you will get a brief introduction to the important episodes in the history of chemistry. Chemistry is one of the oldest sciences and dates back to ancient times.

Early Chemistry in Ancient Civilizations

Before human history was written down, there was no activity that we would have called *chemistry*. The earliest humans used natural materials just as they found them. They wore the skins of the animals they killed. They were hunters and gatherers who ate their food raw. Their tools were stones or the crude shaping of stones into simple, useful implements. They witnessed lightning strikes that resulted in wild fires, but they could not control fire.

Chem Vocab

Bronze is an alloy of copper and tin, not to be confused with brass, which is an alloy of copper and zinc.

Then early men and women learned to harness fire, and their world was never the same. Fire meant cooking, warmth, and metallurgy.

Roll the clock forward many millennia from the Stone Age to the Bronze Age, to the early civilizations of the Greeks, the Chinese, and the Egyptians. The first metal to be used as a structural material was copper, which was used in the making of bronze, an alloy of copper and tin. Bronze is very hard and easily cast and was used to make weapons, utensils, and other household items.

By the 2000s BCE, these early civilizations had begun to distinguish the differences among pure substances that today we would call elements. The elements known to ancient peoples were carbon, iron, copper, silver, gold, mercury, sulfur, tin, lead, and antimony.

Philosophers, especially in Greece, endeavored to understand the fundamental nature of matter. Vigorous debates were held over the question of whether or not matter is infinitely divisible, or if some smallest particle exists that makes up all of matter. Some Greeks, principally Democritus (460?–370?BCE), did believe in such particles and called them atoms, from the Greek word *atomos*, which means "indivisible".

Alchemy during the Middle Ages

From the early centuries of the current era into the Middle Ages, extraordinary gains were made in the understanding of matter. Medieval alchemists learned to synthesize numerous compounds that do not occur in nature—many for their medicinal or therapeutic value, many for economic value, and many just for the sake of discovering new things. People often remember the alchemists largely for their attempts to turn metals like lead into gold. The fact that they failed to do so should not denigrate their many valuable contributions to our modern understanding of chemistry. Alchemists also tried to discover an elixir of life that would have the ability to prolong life and good health, but again without success.

Several new elements were discovered during the Middle Ages: arsenic, zinc, bismuth, and phosphorus. Much credit goes to alchemists in the Middle East who preserved the writings of the ancient Greeks and who continued to make contributions to science during the so-called Dark Ages in Europe. During the Renaissance in Europe, European scientists "rediscovered" the ancient writings and learned of the contributions that had been made by Middle Eastern alchemists.

Figure 2.1: Elements known by 1800

Elements Known by 1800		
Hydrogen	Manganese	Tin
Beryllium	Iron	Antimony
Carbon	Cobalt	Tellurium
Nitrogen	Nickel	Tungsten
Oxygen	Copper	Platinum
Magnesium	Zinc	Gold
Phosphorus	Arsenic	Mercury
Sulfur	Yttrium	Lead
Chlorine	Zirconium	Bismuth
Titanium	Molybdenum	Uranium
Chromium	Silver	

The Chemical Revolution of the Eighteenth Century

Some historians of science consider Robert Boyle (1627–91) to have been the first real chemist in the modern sense because of his insistence on conducting experiments. Boyle discovered the relationship between the pressure and volume of a gas, which is known today as Boyle's law. By 1800 CE, the practice of chemistry had passed from the more mystical studies of the alchemists into the laboratories of men whom we would recognize today as true chemists—Joseph Priestley (1733–1804), Antoine Lavoisier (1743–94), John Dalton (1766–1844), and Humphrey Davy (1778–1829), among others.

Phlogiston

Before the late 1700s air was believed to be a single substance, not a mixture of gases as we know it to be today. Without knowledge of the chemical composition of the gases in air, combustion (burning) was essentially impossible to understand, yet combustion reactions are among our most important chemical reactions.

Alchemists believed that flammable materials contain a substance that they called phlogiston. Phlogiston was believed to be an actual material substance that had mass.

Alchemists said that during combustion, phlogiston is released to the atmosphere. Similarly, alchemists believed that metals contain phlogiston. When a metal is extracted from its ore, phlogiston is gained in the process. Later, if the metal rusts, alchemists believed that phlogiston is released back to the atmosphere.

In the case of combustion reactions, the theory of phlogiston was plausible. When wood burns, it turns into ash. Since the ash clearly weighs less than the wood did, it would seem that something was "given off" by the wood—phlogiston.

In the case of metals, though, the theory made less sense since it would suggest that a metal that now contains phlogiston should weigh more than the original sample of ore that did not contain phlogiston. The reverse is in fact true—a sample of pure metal weighs less than the original ore, not more. This inconsistency, however, seems not to have bothered European alchemists, although alchemists in the Middle East recognized the problem.

Historical Note

Phlogiston was thought to be a volatile component of any material that is combustible and released during the combustion process. Chemists stopped believing in phlogiston after about 1800.

It was the English minister and part-time chemist Joseph Priestley who provided an alternative to the theory of phlogiston. Priestley discovered oxygen gas, although, ironically, Priestly himself still continued to believe in the existence of phlogiston. The role of oxygen in combustion was recognized instead by the French chemist Antoine Lavoisier, who said that during combustion, a fuel combines with oxygen from the atmosphere (instead of giving off phlogiston), producing carbon dioxide and water. Lavoisier also discovered the law of conservation of mass in chemical reactions.

Historical Note

Joseph Priestley was a Unitarian minister in England. In 1791, Priestley's liberal theology and support for the French Revolution forced him and his family to flee England. They settled in Northumberland, Pennsylvania, where Priestley was welcomed by the scientific community of the fledgling United States. Today, Priestley's home is open to the public.

As other gases were discovered—hydrogen, nitrogen, carbon dioxide, and nitrous oxide—chemists began to develop a more correct understanding of the composition of the atmosphere. Because combustion is so central to chemistry, the discovery of oxygen and its role in combustion reactions ushered in a new era of chemistry, making the end of the eighteenth century indeed a revolution in chemistry comparable to the revolution in physics that had occurred in the sixteenth and seventeenth centuries with the work of Nicolas Copernicus, Galileo Galilei, and Isaac Newton.

Besides oxygen, hydrogen, and nitrogen, many more new elements were discovered just before and during the eighteenth century: platinum, cobalt, nickel, magnesium, chlorine, manganese, chromium, molybdenum, tellurium, tungsten, zirconium, uranium, titanium, yttrium, and beryllium.

Historical Note

Antoine Lavoisier may have been one of the greatest chemists who ever lived. Unfortunately, he was also a tax collector at the time of the French Revolution. At the age of fifty, Lavoisier was beheaded on the guillotine. The French mathematician Louis Lagrange lamented, "It took them only an instant to cut off his head, but France may not produce another such head in a century."

The Development of Modern Chemistry during the Nineteenth Century

The first decade of the nineteenth century witnessed the discoveries of many more new elements: vanadium, niobium, tantalum, rhodium, palladium, osmium, iridium, cerium, potassium, sodium, boron, calcium, ruthenium, and barium. Instrumental in this work was the English chemist Humphrey Davy.

Salts that contained sodium and potassium had been known since prehistoric times, but the elements themselves had never been isolated. Davy succeeded not only in isolating sodium and potassium, but calcium, strontium, and barium as well. Chemists were beginning to understand the difference between elements, which cannot be broken down by chemical reactions into simpler substance, and compounds, which can be broken down into simpler substances—the elements they contain.

Modern Atomic Theory

The English chemist John Dalton was responsible for reviving the theory of atoms during the first decade of the nineteenth century. Dalton argued that the law of conservation of mass in chemical reactions and the definite proportions of chemical compounds were convincing evidence for the existence of atoms. In recognition of his work, Dalton is often called the father of modern atomic theory.

With a theory of atoms, chemists began to have a more exact way to define elements and compounds: elements are pure substances that contain only one kind of atom; compounds are pure substances that contain more than one kind of atom and that are present in very definite proportions. Mixtures, on the other hand, were recognized as impure substances that consist of more than one element or compound, but which are present in variable proportions.

Important Point

Elements are pure substances that contain only one kind of atom. Compounds are pure substances that contain more than one kind of atom.

Chem Vocab

The term *periodic* means "having repeated cycles." The properties of elements in the periodic table are repeated at regular intervals.

Laboratory experiments shed light on the chemical composition of compounds. The determination of crude atomic weights of elements gave rise to listing elements in order of weight. But fundamental questions persisted: How many elements are there? Are the chemical and physical properties of elements completely random or does some logical framework exist in which to arrange elements so that their properties make sense? It took more than a century to answer these questions.

The Periodic Table

A major step forward in answering questions like these was made by the Russian chemist Dmitri Mendeleev (1834–1907), who, in 1869, published a periodic table of the elements, arranging the elements in a systematic fashion that, with modifications, is still followed today. With a couple of exceptions, Mendeleev listed the elements in order of increasing atomic weight and then arranged them into rows, or periods, such that elements that fell into the same column, called a family or group, possessed similar chemical and physical properties. By leaving gaps in the periodic table where no known elements appeared to fit, and then estimating what the properties of those elements should be, Mendeleev's table had the great power of prediction.

Within a matter of a few years, other chemists used Mendeleev's predictions to discover several elements that had been unknown in 1869—gallium, germanium, and scandium. These elements fit neatly into the spaces that Mendeleev had left blank, and they had chemical and physical properties that were very close to the ones that Mendeleev had predicted. At the time Mendeleev developed his periodic table, the noble gases had not yet been discovered. Mendeleev lived to witness their discovery and their placement into their own family of elements.

Figure 2.2 – Mendeleev's periodic table

			Ti = 50	Zr = 90	?? = 180
			V = 51	Nb = 94	Ta = 182
			Cr = 52	Mo = 96	W = 186
			Mn = 55	Rh = 104.4	Pt = 197.4
			Fe = 56	Ru = 104.4	Ir = 198
			Ni = Co = 59	Pd = 106.6	Os = 199
H = 1			Cu = 63.4	Ag = 108	Hg = 200
	Be = 9.4	Mg = 24	Zn = 65.2	Cd = 112	
	B = 11	Al = 27.4	?? = 68	Ur = 116	Au = 197
	C = 12	Si = 28	?? = 70	Sn = 118	
	N = 14	P = 31	As = 75	Sb = 122	Bi = 210
	O = 16	S = 32	Se = 79.4	Te = 128	
	F = 19	Cl = 35.5	Br = 80	J = 127	
Li = 7	Na = 23	K = 39	Rb = 85.4	Cs = 133	Tl = 204
		Ca = 40	Sr = 87.6	Ba = 137	Pb = 207
		?? = 45	Ce = 92		
		Er = 56	La = 94		
		Yt = 60	Di = 95		
		In = 75.6	Th = 118		

At first, Mendeleev arranged the known elements in rows instead of in columns. The numbers represent weights relative to hydrogen as measured at that time. Note that the symbols of some elements have changed. The unknown elements for which Mendeleev left blank spaces are Sc (45), Ga (68), Ge (70), and Hf (180).

Historical Note

Probably by coincidence, three elements whose existence was predicted by Mendeleev were all named after countries. Gallium was named after France (*Gaul* in Latin). Germanium was named after Germany. Scandium was named after Scandinavia.

Chemists and physicists still did not know, however, what atoms actually are—just little hard spheres with different masses, or something else? In 1897, the British physicist J. J. Thompson (1856–1940) discovered the electron and concluded that electrons are particles found inside atoms, which meant that atoms could not just be little hard spheres. Scientists realized that the existence of negatively charged particles inside atoms also required the existence of positively charged particles—called protons—to equalize the charges. It required thirty more years—not until 1932— for neutrons to be discovered, an accomplishment made by the British physicist James Chadwick (1891–1974). More will be said about protons, neutrons, and electrons in Chapter 4.

Between 1810 and 1900, a total of thirty-five more new elements were discovered. Notable examples include iodine, silicon, aluminum, fluorine, radium, the noble gases, and most of the rare earths.

The most important figure in the discovery and identification of several of the rare earths was the Finnish chemist Johann Gadolin (1760–1852). Element 64, gadolinium, the first element to be named after a person, was named in honor of Gadolin.

During the 1890s, the English chemist William Ramsey (1852–1916) was responsible for the discoveries of helium, neon, argon, krypton, xenon, and radon. Ramsey was recognized for these achievements by being awarded the 1904 Nobel Prize in chemistry.

Organic Compounds

Chemical compounds are divided into those classified as inorganic substances versus those classified as organic substances. Until the middle of the nineteenth century, chemists thought that organic substances—chemical compounds that contained carbon, hydrogen, and possibly other elements—could be made only by living organisms. That misunderstanding was resolved in 1827, when the German chemist Friedrich Wőhler (1800–82) synthesized the organic compound urea.

Since that time, organic compounds have been considered simply to be substances that contain carbon, hydrogen, and possibly other elements; and inorganic compounds are substances that do not contain both carbon and hydrogen. That recognition launched early chemical industries and ultimately resulted in global chemical companies that are among the world's largest corporations.

Chem Vocab

Chemical substances are divided into two broad classifications: organic compounds and inorganic compounds. Organic compounds may be defined as any substances that contain both carbon and hydrogen (and possibly other elements). Inorganic compounds consist of everything else.

Thermochemistry and Chemical Kinetics

These achievements were not the only ones chemists made during the nineteenth century that resulted eventually in our modern understanding of atoms and molecules. Concurrently, chemists came to understand how elements and compounds either absorb heat from their surroundings or give off heat to their surroundings as chemical reactions occur. These heat effects, summarized in the branch of chemistry known as thermochemistry, are important in understanding the nature of combustion processes and in controlling what products a certain mix of reactants will yield.

The factors that control the rates, or speeds, at which chemical reactions occur, summarized in the branch of chemistry known as chemical kinetics, were discovered, and chemists learned how to control the rates of reactions by altering reactant concentrations, changing the temperature, or introducing a catalyst, a substance that speeds up a chemical reaction without itself being altered in the process.

Chem Vocab

Thermochemistry is the study of the heat effects associated with chemical phenomena.

Chemical kinetics is the study of the rates of chemical reactions and the factors upon which rates depend.

The Discovery of Radioactivity

In 1896, radioactivity was discovered. With that discovery came the realization that atoms are not immutable (unalterable), but that, in fact, they can be transmuted (changed) into atoms of different elements. To change an atom of one element into an atom of a different element requires that the number of protons in the atom be changed. Such a transmutation is a nuclear reaction and is fundamentally different from the ordinary chemical reactions that have been discussed thus far. Ordinary chemical reactions are all about electrons—there are no changes in the numbers of protons or neutrons. On the other hand, nuclear reactions are all about protons and neutrons—the electrons usually are not part of nuclear reactions, although their number can change to remain equal to the number of protons in an electrically neutral atom.

Historical Note

When X-rays and radioactivity were first discovered, scientists did not recognize their danger. Several years elapsed before people began taking safety precautions when taking x-ray photographs or handling radioactive substances.

Chemistry in the Twentieth and Twenty-First Centuries

In 1915, the English physicist Henry Moseley (1887–1915) suggested that a more fundamental property of an element than its atomic weight is the number of protons in atoms of that element, what came to be called the atomic number of an element.

Ordering the elements in the periodic table known at that time according to their atomic numbers instead of atomic weights gave the same order that Mendeleev had decided must be true in order for elements with similar properties to fall into the same columns. Even more importantly, however, before Moseley's work no one knew how many elements there should be with atomic weights between those of lanthanum and hafnium (the so-called lanthanides). By assigning lanthanum atomic number 57 and hafnium atomic number 72, it became clear that there should be fourteen lanthanides. Since thirteen lanthanides were already known, chemists realized there was room for only one more—element 61—lying between neodymium and samarium. Several decades later element 61, promethium, was discovered among the fission fragments in nuclear reactors.

Chem Vocab

The lanthanides are the fourteen elements that follow lanthanum in the periodic table and whose properties are very similar to those of lanthanum. Thirteen of the lanthanides are found in nature. Only promethium (number 61) must be produced artificially.

Quantum Theory

The orderliness of the periodic table, however, begged the question of why elements in the same columns of the periodic table have similar chemical and physical properties. In other words, what was the relationship between an element's atomic number and its properties? What was significant about an element's location in the first column of the table, or the second column, or the sixth column, or whatever column, that imparted to an element certain properties? The answers to those questions were gradually ferreted out by a number of brilliant physicists who, during the first thirty years of the twentieth century, developed quantum theory—the description of matter at the microscopic level of individual atoms and molecules.

Quantum theory demonstrates that chemical properties and everyday chemical reactions are all about electrons. Electrons are arranged in atoms in very specific energy levels. Elements

Historical Note

Henry Moseley was drafted into the British army during World War I. In 1915, Moseley has killed at the battle of Gallipoli at just twenty-seven years of age. After Moseley's death, British scientists were assigned noncombatant duties.

Chem Vocab

Quantum means an indivisible unit of energy.

in the same column of the periodic table have the same number of outermost electrons occupying the same energy levels. Similar configurations of electrons mean similar properties.

Chemical Bonds

With the understanding of the role electrons play in the properties of atoms, chemists came to realize that electrons are the essence of the chemical bonds between atoms. By knowing how many electrons an atom has—and which energy level the electrons occupy—chemists could understand how many bonds a particular kind of atom will form, to what other elements an element will bond, and—most importantly—what structures molecules will assume either in the orderly arrangements of atoms in crystalline solids or in the geometric shapes of molecules in the liquid or gaseous phases.

Just as the arrangement of electrons within atoms determines the properties of those atoms, the arrangement of electrons also determines the shapes of molecules, which in turn determine the properties of those molecules. Once again, chemistry is all about electrons!

The Discovery of Fission and Fusion

Roughly forty years after the 1896 discovery of radioactivity, the nuclear processes of fission and fusion were discovered. Fission and fusion have changed the world forever, although they are a two-edged sword. Peacetime applications of fission include providing an abundant source of energy for Earth's burgeoning population. (Fusion has the same potential, but commercial fusion reactors are not yet technologically possible.) However, as is well known, fission and fusion are also the basis of nuclear weapons, which potentially could destroy Earth's population.

Historical Note

It is a tremendous coincidence of history that the neutron was discovered just prior to Adolf Hitler's ascension to the office of Chancellor of Germany. The discovery of the neutron led to the discovery of nuclear fission, which in turn led to the development of the atom bomb.

Elements beyond Uranium

The discoveries of elements heavier than uranium, the so-called transuranium elements, began in the late 1930s. Prominent in these discoveries was the American chemist Glenn T. Seaborg (1912–99), who received the 1961 Nobel Prize in chemistry for his discovery of plutonium. In the early years of transuranium chemistry, most of the work was carried out under Seaborg's supervision at the University of California at Berkeley. Today, however, active laboratories are also located at Darmstadt, Germany; Dubna, Russia; and several locations in Japan. As of 2012, the periodic table consisted of 118 elements, although the discoveries of four the last six elements are still awaiting confirmation before they can be assigned official names.

Chem Vocab

The elements that lie beyond uranium in the periodic table are called transuranium elements.

Chemistry and Biology

In 1953, James Watson (1928–), Francis Crick (1916–2004), and Maurice Wilkins (1916–2004) solved the mystery of the structure of deoxyribonucleic acid (DNA) and how it serves as the basis of heredity. By the close of the twentieth century, knowledge about DNA and how to manipulate its structure had led to deciphering the genetic codes of living organisms (including humans), genetic engineering, cloning organisms, and genetically modified food crops—all of which have carried their share of controversy. As the twenty-first century dawned, chemists continued to use their understanding of DNA in the fight against cancer, cardiovascular disease, Alzheimer's disease, Parkinson's disease, autism, and communicable infectious diseases—from human immunodeficiency virus (HIV) to the common cold.

Historical Note

University of California chemist Glenn Seaborg was the only person who has ever held a patent on a chemical element. Seaborg held patents on americium and curium.

Conclusion: Chemistry beyond Earth

Chemists are not limited by the boundaries of Earth. Unmanned probes to other planets and their moons have analyzed the chemical compositions of their atmospheres, aqueous and nonaqueous bodies of liquids, and soils.

One of the biggest unanswered questions in science is whether we are alone in the universe. Most scientists are convinced that we are not alone, that extraterrestrial life does exist. One of the major reasons for conducting chemical analyses of other planets and moons is to find out if conditions on those bodies are conducive to life as we know it on Earth.

The pace of discovery in chemistry has not slowed down at all. New discoveries are reported every day. Young people looking for careers would do well to investigate the possibilities and rewards that chemistry has to offer. Great satisfaction awaits the chemists of tomorrow who develop energy sources to replace fossil fuels, who conquer the ravages of disease, or who find ways to feed the world's burgeoning population.

CHAPTER 3

The Language of Science

> ## In This Chapter
>
> ➤ Scientific notation
>
> ➤ The international system (SI) of units
>
> ➤ Derived units
>
> ➤ Temperature scales

Every academic discipline has its own jargon. Chemistry is no exception. The purpose of this chapter is to introduce you to some of the terminology and mathematical notation used in chemistry.

Scientific Notation

Science has its own language that scientists use to communicate with each other. Because scientists may be called upon to measure extremely large numbers (like the distances to far-flung galaxies) and extremely small numbers (like the diameter of a proton), scientific notation is used. Scientists have also agreed to use units of measurement that make communication clearer.

Generally speaking in science, the metric system is preferred, although nonmetric units may persist in areas where they are familiar to scientists working in that area. Most units have some shorthand form of abbreviation that scientists have been using for centuries.

Consider the distance light travels in a vacuum in one year, the distance astronomers call a light year. Traveling at a constant speed of 186,000 miles/second for 31,558,000 seconds (one year), a light year is a distance of 5,869,713,600,000 miles. (Note: A light year is close to that distance, but not exactly that distance. First of all, my calculator only displays 10 digits. Secondly, I used 365-1/4 days as the length of a year, but that is not an exact number either.)

It is cumbersome to have to count digits, but if we did, we could call the distance that light travels in a year approximately 5.87 trillion miles. What scientists usually do, though, is write large numbers like that using scientific notation. A number in scientific notation is written as a nonzero digit, a decimal point, and one or more decimal places followed by "x 10^y," where y is an integer representing powers of 10. Thus, we would write 5,869,713,600,000 miles as 5.87×10^{12} miles. The exponent of 12 is derived by counting the number of digits after the 5 at the beginning of the number. (I could have written more decimal places, but it is my common practice to round numbers off to three digits unless there is reason for more, or fewer, digits.) In general, if we move the decimal point to the left n places, then the power of 10 will be equal to n.

Chem Vocab

Scientific notation refers to writing a number like 3,516 in the form 3.516×10^3, or 0.00052 as 5.2×10^{-4}. In scientific notation, a number is written as a single digit, a decimal point, and any remaining digits times 10 to an exponent.

Now consider the distance between the two hydrogen nuclei in a hydrogen molecule (H_2, or H–H). That distance is 0.000000000064 meter. Again, it is cumbersome to have to count all those zeros, and easy to make a mistake. Therefore, a scientist would express that distance using scientific notation as 6.4×10^{-11} meter. The exponent −11 is derived by counting the number of decimal places up to and including the 6. We obtained a negative exponent this time because we moved the decimal point to the right. If we start with a number that is much less than 1, we will always be moving the decimal point to the right and the power of 10 will always be negative.

Chemists often find themselves in situations in which they need to write very large numbers or very small numbers. An example of a very large number is what is called Avogadro's number, which has a value of 6.02×10^{23}, and is also called a mole. An example of a very small number is the mass of a single atom of copper, which has a value of 1.06×10^{-25} kilogram. Another example of a very small number is the electrical charge on an electron, which has a value of 1.60×10^{-19} coulomb.

Important Point

The number 6.02 x 10^{23} is huge! Written as a decimal number, it would be 602,000,000,000,000,000,000,000, or 602 sextillion, and 6.02 × 10^{23} atoms would be 602 zetta-atoms.

In this book many numbers will have values between 0.001 and 1,000. As a general rule, those numbers will be written in ordinary decimal notation. For numbers smaller than 0.001, or larger than 1,000, however, scientific notation will generally be used.

Units and Their Abbreviations

In the examples given in the preceding section, both British and metric units were used. For example, the mile is a British unit of distance, while the meter is a metric unit; kilogram is metric; units of time, e.g., the "second," are both British and metric. Science, however, tends almost exclusively to use the metric system, and that will be the general practice in this book.

The International System (SI) of Units

By international agreement scientists use almost exclusively the International System of Units (Système International d'Unités, or SI, in French). The fundamental units that are used, and their abbreviations, are the following:

mass	kilogram (kg)	amount of substance	mole (mol)
distance	meter (m)	electric current	ampere (A)
time	second (s)	luminosity	candela (cd)
temperature	Kelvin (K)		

Historical Note

For more than a century the world's standard kilogram has been a cylinder made of 90 percent platinum and 10 percent iridium (by weight) kept in Paris. Beginning in the 1790s, the world's standard meter was defined as one ten-millionth of the distance from the equator to the North Pole. In 1984, the definition of a meter was made more precise—1/299,792,458 the distance that light travels in a vacuum in one second.

Metric Prefixes

It is never wrong to use only the fundamental units. In many situations that just means the numbers are either very large or very small. To keep numbers to reasonable sizes, scientists use units derived from the fundamental set of units using metric prefixes. These prefixes and their abbreviations are the following:

femto (f)	10^{-15}	centi (c)	10^{-2}
pico (p)	10^{-12}	deci (d)	10^{-1}
nano (n)	10^{-9}	kilo (k)	10^{3}
*micro (μ)	10^{-6}	mega (M)	10^{6}
*milli (m)	10^{-3}	giga (G)	10^{9}

* The symbol "μ" is the Greek letter "mu" and is used because m was already in use. The letter m can be the abbreviation for either meter or milli. It is usually clear from the context which meaning is intended.

A mass of 0.000005 kg can also be written as 5 μg. A distance of 6,000 m can also be written as 6 km, or 6.0 km, or 6.00 km, depending upon how precisely the distance is known.

Historical Note

The origin of the prefix *femto* is the Danish word *femten*, meaning "fifteen". *Pico* is derived from the Italian word *piccolo*, meaning "small." *Nano* is derived from the Greek word νᾶνος, meaning "dwarf."

Derived Units

In this book you will encounter other units, such as the newton, a unit of force; the joule, a unit of energy; and the pascal, a unit of pressure. These are called derived units. A newton (N) is the same as a kg·m/s². A joule (J) is the same as a newton·meter (N·m), or a kg·m²/s². A pascal (Pa) is the same as a newton per square meter (N/m²). Derived units are compositions of SI units.

Rather trivial examples of derived units also include units for area (cm²), volume (cm³), velocity (m/s), and density (g/cm³).

Nonmetric Units

There are units that have been in use for so many years that scientists continue to use them even though they are not SI units. These units, their abbreviations, and their equivalencies to SI units are the following:

Length	Ångstrom	(Å)	$1 \text{ Å} = 10^{-10} \text{ m}$
Energy	calorie	(cal)	$1 \text{ cal} = 4.184 \text{ J}$
Energy	electron volt	(eV)	$1 \text{ eV} = 1.60 \times 10^{-19} \text{ J}$
Pressure	atmosphere	(atm)	$1 \text{ atm} = 760 \text{ mm Hg}$
Pressure	torr	(torr)	$1 \text{ torr} = 1 \text{ mm Hg}$
Temperature	degrees Celsius	(°C)	$T \text{ (°C)} = T(K) - 273.15$
Volume	liter	(L)	$1 \text{ L} = 1000 \text{ cm}^3$

Historical Note

The unit Ångstrom is named after the Swedish physicist Anders Jonas Ångström (1814–74). In Swedish, the letter *A* is always written Å. In English, the circle over the *A* is often omitted.

Temperature Scales

There are several temperature scales in existence. Canadian readers are accustomed to using the Celsius scale (°C). Readers who live in the United States are more familiar with the

Fahrenheit scale, (°F). Equivalent temperatures on the two scales include the following:

Boiling point of water (at 1 atm)	100°C	212°F
Freezing point of water (at 1 atm)	0°C	32°F

A little arithmetic shows that the difference between the boiling point and the freezing point of water is 100° on the Celsius scale and 180° on the Fahrenheit scale. Therefore, the magnitude of a Celsius degree is $180 \div 100 = 1.8$ times the size of a Fahrenheit degree. In addition, we see that the two scales are offset by 32°, the difference between the two values of the freezing point of water. Thus, the equation that relates the two temperature scales is the following:

$$T \,(°F) = 1.8 \, T \,(°C) + 32°$$

For example, if T = 21°C (a pleasant spring or fall day), on the Fahrenheit scale the temperature would be 70°F. If T = −10°F (a cold northern winter day), on the Celsius scale the temperature would be −23°C.

Absolute Temperature Scales

Fahrenheit and Celsius are not absolute scales of measurement in the sense that they were chosen somewhat arbitrarily. Gabriel Fahrenheit (1686–1736) chose as 0° the coldest temperature he could get in his laboratory. When he marked off the degrees on his thermometer, water boiled at 212°. When Anders Celsius (1701–44) invented his temperature scale, he chose the freezing point of water to be 0° and water's boiling point to be 100°.

Historical Note

Because of the difference of 100 degrees, for many years Celsius's temperature scale was called the centigrade scale, indicating 100 divisions. Only relatively recently was the decision made to refer to all temperature scales by their inventors.

The problem with these rather arbitrary decisions is that zero on any measurement scale should indicate the complete absence of whatever it is that is being measured. For example, a ruler or meter stick measures length. Zero truly means zero length or the complete absence of length. A bathroom scale measures weight. Zero truly means the complete absence of weight. Temperature measures how hot an object is. Zero degrees Celsius or Fahrenheit, however, does not mean the complete absence of heat. If they did, when the temperature at night drops to 0°C or 0°F, then the atmosphere would have no heat content and the atmosphere would collapse. Clearly, the atmosphere does not do that.

The solution to this problem lead to the creation of absolute temperature scales, ones on which 0° truly means the absence of any heat. The first absolute temperature scale was the Kelvin scale, named for the British physicist William Thomson (1824–1907), who went by the title Lord Kelvin. It is perhaps a technical point, but absolute temperature scales do not use the degree symbol; hence, a temperature on the Kelvin scale is expressed in kelvins (k), not degrees Kelvin, and not capitalized. The size of a degree on the Kelvin scale is the same as the size of a degree on the Celsius scale, so the conversion between the two scales is relatively simple:

$$T (k) = T (°C) + 273.15$$

On the Kelvin scale the freezing point of water is 273.15 k and the boiling point is 373.15 k.

A similar absolute scale is based on the Fahrenheit scale, and is named for the Scottish physicist William Rankine (1820–72). On the Rankine scale water freezes at 491.69 degrees and boils at 671.69 degrees.

Absolute Zero

In theory there really is a coldest temperature possible in the universe, although in practice it is impossible to actually reach that temperature. To cool something down, an object has to be able to give off heat to something else that is colder than it is. Since there is nothing colder than zero kelvin in the universe, there is nowhere left to which to transfer that last miniscule bit of heat. Scientists have reached temperatures on the order of one-millionth of a degree above absolute zero, but not absolute zero itself. For the sake of discussion, however, absolute zero corresponds to −273.15°C, or −459.67°F.

Conclusion

If you were not familiar already with the units and vocabulary introduced in this chapter, don't worry. Since we will be referring to them repeatedly throughout this book, you should find that they eventually will become second nature to you.

CHAPTER 4

Elements and Compounds

In This Chapter

➤ Atoms

➤ Elements

➤ Compounds

Pure substances consist of elements and compounds. Elements are pure substances that contain only one kind of atom. Compounds are pure substances that contain more than one kind of atom. In this chapter you will learn what atoms of made of, what makes different kinds of atoms different from each other, and how these differences give us the 118 elements that have been discovered as of 2012.

The Composition of Atoms

Atoms are the fundamental building blocks of matter. Atoms link together to make molecules by an attraction that chemists call the chemical bond. If molecules contain atoms of different elements in fixed ratios, then they are called compounds. Mixtures consist of two or more elements or compounds in variable ratios.

Chem Vocab

Atoms are fundamental substances that cannot be further subdivided by chemical processes.

Compounds are fundamental substances that can be subdivided into their component atoms by chemical processes.

An atom consists of the following three smaller particles:

➤ Protons

➤ Neutrons

➤ Electrons

Protons and neutrons are located in a very tiny center of the atom called the nucleus, a term that was borrowed from biologists' definition of the region in the center of a cell of a living organism. Protons possess a positive electrical charge, whereas neutrons are electrically neutral. Protons and neutrons are relatively massive particles, with neutrons being just slightly heavier than protons Electrons possess a negative electrical charge and have only about $\frac{1}{1,840}$ the mass of a proton or neutron. Electrons are located in the relatively empty region of space surrounding the nucleus.

Table 4.1: Components of Atoms

Chem Vocab

The nucleus is the center of an atom. It consists of two particles—positively charged protons and electrically neutral neutrons. The negatively charged electrons are located outside the nucleus.

Kind of Particle	Charge	Relative Mass	Location
Proton	+	1836	Nucleus
Neutron	0	1839	Nucleus
Electron	–	1	Outside Nucleus

Although of little consequence in chemical reactions, nucleons (protons and neutrons) are not actually fundamental particles, but are themselves composed of even smaller particles called quarks. There are various kinds of quarks, but the two kinds that make up nucleons are called up quarks, which have a charge of +2/3, and down quarks, which have a charge of -1/3. A proton is made up of two up quarks and one down quark for a total charge of +1. A neutron is made up of one up quark and two down quarks for a total charge of 0

As such, protons and neutrons possess measurable diameters (about 10^{-15} m) and occupy a volume that determines the volume of an atomic nucleus. Electrons, however, are indeed fundamental particles with no measurable dimensions and no known internal structure.

Atoms are incredibly tiny particles. About 100,000,000 atoms can be placed in a row 1 centimeter in length. But even then, an atom is mostly empty space, with its nucleus occupying only about 10^{-12} the volume of the atom. To get a feeling for this, picture a football stadium as representing the size of the entire atom. To the same scale, the nucleus would be about the size of a pea located at the center of the 50-yard line! The electrons would be

Chem Vocab

Quarks are the fundamental particles of which protons and neutrons are made. The word *quark* was coined by the American physicist Murray Gell-Mann who, in his own words, made that choice when "in one of my occasional perusals of *Finnegans Wake*, by James Joyce, I came across the word 'quark' in the phrase 'Three quarks for Muster Mark.'"

inside the rest of the stadium, but they are so infinitesimally small that the stadium would look essentially empty.

Chemical Elements

An element is defined as a pure substance that contains only one kind of atom. The phrase *kind of atom* refers to the number of protons that an atom has. Since protons come in whole units only, whole numbers starting with 1 can be used to identify elements. Element number one is hydrogen, with one proton. Element number two is helium, with two protons. The heaviest naturally occurring element is uranium, with ninety-two protons. Elements heavier than uranium have been produced artificially, with the heaviest known element (as of 2012) being number 118, which has not yet been named.

Under standard conditions (1 atmosphere pressure, 25°C) elements can exist as solids, liquids, or gases. Most metals and metalloids, and several nonmetals are solids under standard conditions. A few elements, like mercury (Hg) and bromine (Br), are liquids.

Chem Vocab

An element is a pure substance that contains only one kind of atom. An atom's defining property is its number of protons. Change the number of protons and it is a different element.

Some elements are gases and include the following:

➤ Hydrogen (H)

➤ Nitrogen (N)

➤ Oxygen (O)

➤ Fluorine (F)

➤ Chlorine (Cl),

➤ The noble gases helium, neon, argon, krypton, xenon, and radon.

Matter can be divided into smaller and smaller pieces. A molecule is the smallest unit of a pure substance that retains the properties of that substance. For example, the nitrogen and oxygen gases that make up most of the atmosphere are in the forms N_2 and O_2, respectively.

Important Point

Just from appearance alone, it may not be obvious whether something is a pure substance or a mixture. Air and pond water look like pure substances, but they are mixtures. Nor may it be obvious whether something is an element or a compound. Gold is an element but fool's gold is a compound composed of iron and sulfur, and is completely unrelated chemically to gold.

Chem Vocab

A molecule is the smallest unit of a pure substance that retains the properties of that substance.

We refer to these as nitrogen and oxygen molecules. Sometimes molecules are only single atoms, as in the case of the noble gases, which in pure form are He, Ne, Ar, Kr, Xe, and Rn.

One way to picture molecules as being the smallest units that retain a substance's properties is to use an analogy—eggs in a carton. A carton usually contains a dozen eggs, but it could contain two dozen or only half a dozen; it does not matter for this example. If we start with two dozen eggs, we can divide them into two sets of one dozen eggs each. In turn, we can divide one dozen eggs into two sets of six eggs each. They are still eggs and individually have all the properties that eggs have. We can divide the six eggs into two sets of three eggs each. We can divide three eggs into two sets, one with two eggs and one with only a single egg.

Once we have reached that single egg, the division stops. We cannot divide an egg into parts and still call it an egg. Only a whole egg has the properties of eggs, not the yolk alone, or the albumin, or any other pieces. A molecule is the same way. A molecule may contain more than one atom (piece), but we cannot divide a whole molecule any further and retain the properties of that molecule. Water contains hydrogen and oxygen. If we split water into hydrogen and oxygen, it is not water anymore. In fact, the properties of hydrogen gas and oxygen gas do not even closely resemble the properties of water.

Allotropes

Some elements can exist in different physical forms called allotropes (*allo* means "different," *trop* means "form"). Different forms of the same element may have distinctly different chemical and physical properties. For example, carbon has several allotropes that include graphite and diamonds. Oxygen exists in the atmosphere mostly as diatomic molecules, O_2, but can also exist as the triatomic molecule ozone, O_3. Ozone is significantly more reactive than O_2; whereas O_2 is essential to life, O_3 is harmful. Elements that are solids under standard conditions may have allotropes that differ primarily in their crystalline structures.

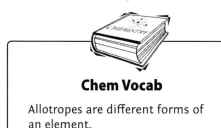

Chem Vocab

Allotropes are different forms of an element.

Isotopes

Unlike the number of protons in atoms of an element, the number of neutrons is not fixed for a particular element. Neutrons have an important function in atoms: they provide the glue that holds nuclei together.

Remember that protons all possess a positive charge. Charges with the same sign (that is, both are positive or both are negative) tend to repel each other. Without neutrons, no atom could exist that had two or more protons. By means of an extremely strong force of nature called the strong nuclear force, protons and neutrons are attracted to each other, which effectively holds multiple protons together so they don't fly apart. Starting with helium, all atoms have to have one or more neutrons. Only hydrogen atoms—and even then only one isotope of

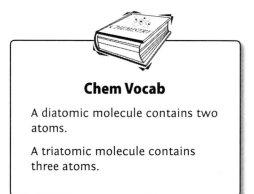

Chem Vocab

A diatomic molecule contains two atoms.

A triatomic molecule contains three atoms.

Periodic Table of the Elements

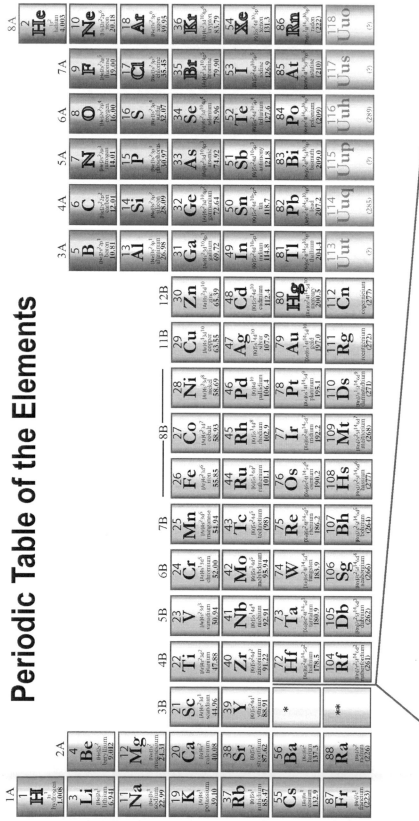

Lanthanide Series*

Actinide Series**

Los Alamos National Laboratory Chemistry Division

© Copyright 2011 Los Alamos National Security, LLC. All rights reserved. The public may copy and use this information without charge, provided that this Notice and any statement of authorship are reproduced on all copies. Neither the Government nor LANS makes any warranty, express or implied, or assumes any liability or responsibility for the use of this information.

Figure 4.1 – Periodic table

Chem Vocab

Isotopes are atoms of the same element that have different numbers of neutrons.

hydrogen—can exist without containing neutrons. A hydrogen atom contains only one proton, so there is no repulsive force in the nucleus.

Although an atom must have neutrons, the number of neutrons it has still is not a fixed quantity. Atoms of the same element can have a different number of neutrons. We call these atoms isotopes of that element. *Iso* means "same," *tope* means "place," just like the use of *tope* in the term *topographical map*.

Isotopes occupy the same place, or location, in the periodic table of the elements (figure 4.1).

For a given number of protons, there usually is a small range of numbers of neutrons that the stable isotopes of an element will have, where *stable* refers to whether or not the isotope is radioactive. Through calcium (element twenty), the number of neutrons in atoms is about equal to the number of protons. After calcium the number of neutrons begins to increase, mainly because the repulsive forces between so many protons is beginning to become quite large. Bismuth (element eighty-three) is the heaviest element to possess stable isotopes. Beginning with polonium (element eighty-four), none of the heaviest elements has any stable isotopes. All of the isotopes of the heaviest elements are radioactive.

The number of electrons in an electrically neutral atom is easy to determine—it is equal to the number of protons. Protons and electrons have the same magnitude of charge, just opposite signs, so equal numbers of protons and electrons will give an atom a net charge of zero, which is what *electrically neutral* means.

Scientists have a convenient symbolism that is used to represent particular isotopes of an element X. That symbol is $_Z^A X$, where the subscript Z is the number of protons and is called the atomic number of the element. The superscript A is the sum of the numbers of protons and neutrons and is called the mass number of a particular isotope. Since the identity of an element is determined by its number of protons, the atomic number and the symbol must match. To find the number of neutrons in an isotope, simply subtract the atomic number from the mass number.

Chem Vocab

Radioactive means that an atom is unstable. Eventually the atom will disintegrate, emitting either a small particle or short-wave radiation.

Table 4.2: Examples of Isotopes of Elements

Chem Vocab

The atomic number of an element is equal to the number of protons in the atoms of that element.

The mass number of an isotope is equal to the sum of the numbers of protons and neutrons in the atoms of that isotope. Isotopes of an element must have the same atomic number, but different mass numbers.

Element	Stable	Stable	Radioactive
Hydrogen	$^{1}_{1}H$	$^{2}_{1}H$ (deuterium)	$^{3}_{1}H$ (tritium)
Carbon	$^{12}_{6}C$	$^{13}_{6}C$	$^{14}_{6}C$
Cobalt	$^{79}_{35}Br$	$^{81}_{35}Br$	$^{82}_{35}Br$
Uranium	none	none	$^{238}_{92}U$

Chemical Compounds

Compounds are pure substances that contain more than one element combined in definite, or fixed, proportions by mass (as opposed to mixtures, in which the proportions of the different elements are variable).

With the exception of only a few elements such as nitrogen, oxygen, helium, sulfur, and gold, elements usually are not found in pure form but are combined with other elements in compounds. Formulas for compounds list the elements that are present, with the relative numbers of atoms designated as subscripts. For example, a familiar compound is water (H_2O), which contains hydrogen and oxygen. Other examples include carbon dioxide (CO_2), propane (C_3H_8), ethyl alcohol (C_2H_6O), sulfuric acid (H_2SO_4), and many common minerals——galena (lead sulfide, PbS), hematite (an iron oxide, Fe_2O_3), malachite (a copper carbonate, $Cu_2CO_3[OH]_2$), rock salt (sodium chloride, NaCl), limestone (calcium carbonate, $CaCO_3$), gypsum (calcium sulfate, $CaSO_4$), quartz (silicon dioxide, SiO_2), and many, many more.

Compounds exhibit two fundamental laws of chemistry: the law of definite proportions, and the law of multiple proportions. The law of definite proportions says that the relative masses of elements in a particular compound are in a fixed ratio. Thus, in water, H_2O, the ratio by

Important Point

Compounds may be separated into their elements by ordinary chemical reactions. An element, however, cannot be changed by an ordinary chemical reaction. To change an element requires a nuclear reaction.

mass of oxygen to hydrogen is 8 to 1. In carbon dioxide, CO_2, the ratio of oxygen to carbon is 8 to 3. In lead sulfide, PbS, the ratio of lead to sulfur is 207 to 32. Typically what happens in a chemical reaction is that compounds exchange atoms to form new compounds. There are over 61 million known chemical compounds in the world. Each compound has a unique formula and its own name.

The law of multiple proportions states that elements can combine in different ratios to give different compounds. Compare water (H_2O) and hydrogen peroxide (H_2O_2). Both compounds contain the elements hydrogen and oxygen, but in different proportions. In H_2O, the proportion of oxygen to hydrogen is 8 to 1, but in H_2O_2, the proportion is 16 to

Important Point

A chemical compound can go by several names, but it can have only one formula. Change the formula of the compound, and the name changes.

1. A different pair of compounds consists of carbon monoxide (CO) and carbon dioxide (CO_2). In CO the ratio of oxygen to carbon is 4 to 3; in CO_2 the ratio is 8 to 3. Carbon monoxide and carbon dioxide both contain carbon and oxygen, but they are distinctly different substances with very different chemical and physical properties—carbon monoxide

is extremely toxic, whereas we inhale and exhale carbon dioxide with every breath.

Sometimes the number of compounds that two or more elements can form can be staggering. Carbon and hydrogen, for example, combine to form a class of organic compounds called hydrocarbons. Hydrocarbons can have general formulas of C_nH_{2n+2}, C_nH_{2n}, C_nH_n, and C_nH_{2n-2}, as well as other combinations. Compounds with the general formula C_nH_{2n+2} include methane (CH_4), butane (C_4H_{10}), and octane (C_8H_{18}) since n can be virtually any whole number. Compounds with the general formula C_nH_{2n} include ethylene (C_2H_4), propylene (C_3H_6), and cyclohexane (C_6H_{12}). Examples of compounds with the general formula C_nH_n are benzene (C_6H_6) and acetylene (C_2H_2), although because of its structural properties, acetylene can also be classified as a C_nH_{2n-2} compound.

Isomers

Chem Vocab

Isomers are chemical compounds that have the same molecular formulas, but different structural formulas. The atoms in different isomers are connected together differently.

The reason there is such a huge diversity of chemical compounds in the world is because of isomerism. Isomers are compounds that have the same molecular formula but different structures, or ways in which their atoms are connected.

For example, there are three compounds called pentanes. They all have the molecular formula C_5H_{12}, but they have different structural formulas:

		H		CH$_3$
		|		|
H$_3$C–CH$_2$–CH$_2$–CH$_2$–CH$_3$	H$_3$C–CH$_2$—C–CH$_3$		H$_3$C–C–CH$_3$
		|		|
		CH$_3$		CH$_3$

n-pentane	isopentane	neopentane
b.p: 36˚C (97˚F)	28˚C (82˚F)	9.5˚C (49˚F)

The abbreviation *b.p.* means "boiling point." The *n* in n-pentane stands for "normal." *Normal* just means it is the straight-chain isomer.

Having different structures means these are different compounds, and they have different chemical and physical properties. For example, each of these compounds has a different boiling point, as shown. Ethanol (also called ethyl alcohol) and dimethyl ether are isomers because they both have the molecular formula C_2H_6O. However, their structures are very different, and therefore their properties are very different:

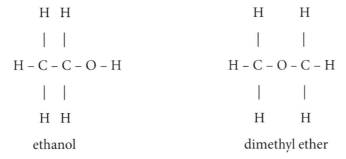

Alcohols contain a C–O–H structure, with the oxygen atom lying between a carbon atom and a hydrogen atom. Ethers contain a C–O–C structure, with the oxygen atom lying between two carbons atoms. The difference in structure explains different properties such as volatility (how easily they evaporate), boiling points, and solubility in common solvents.

Also, you probably recognize ethanol as the alcohol found in alcoholic beverages, but you would never drink ether. You could inhale ether's vapors, and they would probably put you to sleep. In fact, before the development of modern anesthetics, patients were often put to sleep before surgery by inhaling diethyl ether. The practice was unsafe enough that it has been discontinued in most parts of the world, but diethyl ether is still sold as starting fluid for automobiles on cold days.

Sometimes isomers can have structural formulas that include straight chains, branched chains, and rings. Consider compounds with the molecular formula C_6H_{12}. C_6H_{12} has a large number of isomers, although only a few are shown below:

$$H_2C=CH-CH_2-CH_2-CH_2-CH_3 \qquad\qquad H_3C-HC=CH-CH_2-CH_2-CH_3$$

1   2

```
                                                          H  H
                                                          |  |
   CH₃  CH₃               CH₂ – CH₂                `H – C – C – CH₃
    |    |               /         \                      |  |
 CH₃–C = C–CH₃          H₂C         CH₂            H – C – C – CH₃
                         \         /                      |  |
                          CH₂ – CH₂                       H  H

     3                      4                             5
```

The names of these compounds are the following:

1. 1-Hexene
2. 2-Hexene
3. 2,3-dimethyl-2-butene
4. Cylohexane
5. 1,2-Dimethylcyclobutane

Straight-chain or branched chain compounds that contain a C=C double bond are different kinds of hexanes. A hexagon-shaped ring of C_6H_{12} is called cyclohexane. Rings with the formula C_6H_{12} do not contain a double bond. All of the different compounds of C_6H_{12} have their own chemical and physical properties.

Conclusion

Chemistry is the study of matter and its transformations. Matter is made of elements and compounds. Throughout this book we will be studying the structure of matter at the level of atoms and molecules, as well as the properties of matter at the macroscopic level of solids, liquids, gases, and mixtures.

CHAPTER 5

Mixtures of Substances

In This Chapter

➤ The properties of mixtures

➤ Solutions

➤ Units of concentration of solutions

Only rarely do we encounter substances in nature in the form of pure elements or compounds. Most of the time we encounter substances that have been mixed together. Usually a mixture is not uniform in composition and is referred to as being heterogeneous. If the mixture is uniform in composition, however, it is called a solution and is referred to as being homogeneous. This chapter is about both heterogeneous and homogeneous mixtures.

Mixtures and Their Properties

Mixtures contain more than one element, but in variable, not fixed, proportions. We refer to the different elements or compounds in a mixture as the mixture's components. The component, or phase, present in greatest amount is called the solvent, and any other components are called solutes. The fact that the components are not present in fixed proportions is what distinguishes a mixture from a compound, because the elements that make up a compound *are* present in fixed proportions.

Chem Vocab

A compound contains two or more elements in fixed proportions; the relative amounts of the elements are the same regardless of the source of the sample.

In a mixture there also are two or more elements, but not in any fixed proportions; the relative amounts of the components of the mixture can vary from sample to sample.

There are two kinds of mixtures: homogenous mixtures and heterogeneous mixtures. Another name for a homogeneous mixture is a solution. The term *homogeneous* means the solution is the same throughout; any solutes are dispersed evenly throughout the solvent. In heterogeneous mixtures, the various components are not distributed evenly.

Many times the different components, or phases, may in fact be visible. Simple examples would be sand and gravel, salt and pepper, and oil and water. Grains of sand are much finer than pieces of gravel, so the difference between them is easy to see. Salt and pepper are different colors. Oil is less dense than water and floats on water. In each case we can see the different components of the mixtures with the naked eye.

Chem Vocab

Homogeneous means "uniform in composition."

Heterogeneous means "different throughout."

Sometimes the different components may be too small to be seen with the naked eye but can be distinguished when viewed under a microscope. As long as the components are not evenly distributed, it is still a heterogeneous mixture.

Generally speaking, heterogeneous mixtures themselves do not have distinguishing properties of interest. Only the individual components have properties that define them. Because the components in homogeneous mixtures (solutions) are uniformly mixed together, properties of the mixtures themselves may be important. Therefore, most of our discussion of mixtures will deal with solutions.

Ways to Separate the Components of Mixtures

An important property of compounds is that chemical processes are required to separate the elements that make up a compound. Mixtures are different. The components of mixtures can be separated using physical processes. Examples of physical processes include the following:

➤ Fractional distillation, in which any liquid solutes or solvents can be boiled off and collected separately, leaving solid solutes behind

➤ Chromatography, in which mixtures of liquids or gases can be separated into their individual constituents using differences in molecular weight or molecular structure

➤ Filtration, in which a liquid is poured through a filter paper, leaving any solids behind on the paper

➤ Decantation, in which the solid portions of a mixture are allowed to settle to the bottom of a container, and the liquid portion is poured into another container

➤ Gaseous diffusion, in which mixtures of gases are separated by making use of the differences in speed in which gases of different weights travel through a container

➤ Mechanical separation, such as picking bits of gravel out of a pile of sand or using tweezers to separate salt crystals from pieces of pepper.

Important Point

A mixture does not require a chemical reaction to separate its components. The components of a mixture can be isolated by physical processes. Cooling air sufficiently causes each gas to liquefy at a particular temperature. Evaporating salt water removes the water and leaves the salts behind. Tweezers can be used to separate sand from gravel, or salt crystals from pepper.

The Variety of Solutions

With most of our familiar solutions, water is the solvent and the solution is called an aqueous solution. Milk, juices, soft drinks, alcoholic drinks, and sports beverages are all aqueous solutions. Ponds, lakes, and oceans are all aqueous solutions in which the solutes are dissolved solids, gases, and various organic compounds. The main component of blood is water, so blood is an aqueous solution.

Gasoline is a homogeneous mixture of many different compounds in which water is not the solvent—in fact, one usually takes care that gasoline is not contaminated with water. Paints are solutions in which water may or may not be the solvent. In oil-based paints oil is the solvent.

Air is also a solution. In air, the gas present in greatest quantity is nitrogen, so N_2 is the solvent. All other gases—mostly oxygen (O_2), argon (Ar), carbon dioxide (CO_2), and water vapor—are the solutes.

Chem Vocab

Aqueous means "containing, or dissolved in, water." In an aqueous solution, water is the solvent.

Important Point

By number of molecules, the atmosphere is 78 percent N_2, 21 percent O_2 and 1 percent Ar, with only trace amounts of all other gases.

Units of Concentration for Solutions

The concentration of a solution is meant to express how much solute is present compared to the amount of solvent. There are several different units of concentration used in science; each one is preferred in particular applications. The units of concentration we will use are the following:

➤ Percent by weight (or mass)

➤ Percent by volume

➤ Molarity

➤ Mole fraction

➤ Molality

Percent by Weight

A glucose solution administered intravenously to a hospital patient is usually labeled as the percent glucose by weight. For example, the solution could be 10 percent glucose in water; every 100 g of total solution contains 10 grams of glucose and 90 grams of water. Such solutions are simple to prepare and the concentration as a percent by weight does not change if, for example, the temperature changes. In certain types of chemical calculations, however, concentrations of solutes need to be included and percent by weight is not very convenient for calculations. So, we might label a solution that way, but probably not if the solution is going to be further diluted or used in a chemical reaction.

Important Point

Expressing concentrations as percent by weight is common with aqueous solutions. Using percent by volume is common with mixes of gases.

Percent by Volume

Percent by volume is the most common unit of concentration used for mixtures of gases. It is also true that the percent by relative numbers of molecules in a mixture of gases has the same value as the percent by volume of those gases. Therefore, when we say that the atmosphere is 78 percent N_2 by volume, we mean that in a sample of 100 liters of air we can think of 78 liters of the sample as being N_2. However, we can also say that for every 100 molecules of air, 78 of the molecules are N_2.

Because concentrations as relative numbers of particles are so common in mixtures of gases, related units are the following:

> ➤ Parts per million, or ppm

> ➤ Parts per billion, or ppb

> ➤ Parts per trillion, or ppt

Thus, we could also describe the concentration of N_2 in the atmosphere as 780 thousand parts per million. It is more common, however, to use *ppm* for gases that are present in very small quantities. For example, the concentration of carbon monoxide (CO) in a polluted atmosphere might be 10–20 ppm, or 10–20 molecules of CO for every million air molecules total.

Molarity

The molarity of a solution (symbolized *M*) is defined as the number of moles of solute present per liter of total solution. For example, if 2.00 moles of table salt are dissolved in enough water to make one liter of solution, then the molarity of the solution is 2.00 M, pronounced "2.00 molar." Molarity is probably the most common unit of concentration used in chemistry, especially when working with aqueous solutions.

There are, however, disadvantages to using molarity. First of all, the volume of a liquid solution changes with temperature—usually liquids expand as temperature increases and contract as temperature decreases. Molarity, therefore, changes as temperature changes. Fortunately, the volume change can usually be neglected except in the most precise work.

Chem Vocab

Molarity means "moles of solute per liter of solution."

Mole fraction means "moles of one component of a solution per total number of moles of all components of the solution."

Molality means "moles of solute per kilogram of solvent."

Second, notice that molarity is defined in terms of volume of total solution, not just volume of solvent.

In the example above, if you want the final volume to be 1.00 liter, you cannot just add 2.00 moles of table salt (117 grams of salt) to 1.00 liter of water. When solutions are prepared, some expansion or contraction, however slight, will occur, so the final volume is not exactly 1.00 liter. To prepare the solution, therefore, it is best to dissolve the table salt completely in a slightly smaller volume of water—80–90 mL for example—and then dilute the resultant solution to a final volume of 1.00 liter.

Because the use of molarity simplifies many calculations, especially in thermodynamics and kinetics, the advantages of using molarity outweigh the disadvantages. In this book we will mostly be using molarity to express concentrations of aqueous solutions. In other situations, though, other sets of units may be more convenient.

Mole Fraction

Suppose a solution has two components, which we will label A and B. The mole fraction (symbolized by the Greek letter *chi*, χ) of component A is defined as the number of moles of A present divided by the total number of moles present:

$$\chi_A = \frac{\text{number of moles of component A}}{\text{total number of moles}}.$$

Notice that mole fraction is dimensionless—moles in the denominator cancel moles in the numerator. We may speak of the mole fraction of either the solute or the solvent, or, in the case of several solutes, the mole fraction of each solute.

The use of mole fraction has several advantages. One advantage is that the number of moles of a component of a solution is determined by the mass of that component. Mass does not change with temperature, so mole fractions avoid any problems of being temperature dependent. Another advantage is that the masses of the various components do not change when the components are mixed together, so preparing a solution of known mole fraction is relatively simple—just weigh each component and then mix them all together.

We will use mole fraction in a couple of places in this book: one in relation to gas laws and the other in relation to the effect on the vapor pressure of a liquid caused by dissolving a solute in it.

Molality

The molality of a solution (symbolized m, and not to be confused with molarity, symbolized with a capital M) is defined as the number of moles of solute per kilogram of solvent. Molality has the same advantage as mole fraction in that it is temperature independent.

A solution with a specified molality is also easy to prepare because all we have to do is weigh out a certain amount of solute and a certain amount of solvent and then mix the two together.

Molality is useful in two kinds of problems in particular—determining the effects on boiling points and melting points of dissolving a solute in a solvent. We will encounter these phenomena in the Chapter 23.

Conclusion

A lot of new vocabulary was introduced in this chapter. Most of it may have been new to you and may seem like a lot to remember. With time, however, you should become more familiar with these terms.

Periodic Table of the Elements

1A	2A	3B	4B	5B	6B	7B		8B		11B	12B	3A	4A	5A	6A	7A	8A
1 **H** 1s^1 hydrogen 1.008																	2 **He** 1s^2 helium 4.003
3 **Li** [He]2s^1 lithium 6.941	4 **Be** [He]2s^2 beryllium 9.012											5 **B** [He]2s^{2}2p^1 boron 10.81	6 **C** [He]2s^{2}2p^2 carbon 12.01	7 **N** [He]2s^{2}2p^3 nitrogen 14.01	8 **O** [He]2s^{2}2p^4 oxygen 16.00	9 **F** [He]2s^{2}2p^5 fluorine 19.00	10 **Ne** [He]2s^{2}2p^6 neon 20.18
11 **Na** [Ne]3s^1 sodium 22.99	12 **Mg** [Ne]3s^2 magnesium 24.31											13 **Al** [Ne]3s^{2}3p^1 aluminum 26.98	14 **Si** [Ne]3s^{2}3p^2 silicon 28.09	15 **P** [Ne]3s^{2}3p^3 phosphorus 30.97	16 **S** [Ne]3s^{2}3p^4 sulfur 32.07	17 **Cl** [Ne]3s^{2}3p^5 chlorine 35.45	18 **Ar** [Ne]3s^{2}3p^6 argon 39.95
19 **K** [Ar]4s^1 potassium 39.10	20 **Ca** [Ar]4s^2 calcium 40.08	21 **Sc** [Ar]4s^{2}3d^1 scandium 44.96	22 **Ti** [Ar]4s^{2}3d^2 titanium 47.88	23 **V** [Ar]4s^{2}3d^3 vanadium 50.94	24 **Cr** [Ar]4s^{1}3d^5 chromium 52.00	25 **Mn** [Ar]4s^{2}3d^5 manganese 54.94	26 **Fe** [Ar]4s^{2}3d^6 iron 55.85	27 **Co** [Ar]4s^{2}3d^7 cobalt 58.93	28 **Ni** [Ar]4s^{2}3d^8 nickel 58.69	29 **Cu** [Ar]4s^{1}3d^{10} copper 63.55	30 **Zn** [Ar]4s^{2}3d^{10} zinc 65.39	31 **Ga** [Ar]4s^{2}3d^{10}4p^1 gallium 69.72	32 **Ge** [Ar]4s^{2}3d^{10}4p^2 germanium 72.64	33 **As** [Ar]4s^{2}3d^{10}4p^3 arsenic 74.92	34 **Se** [Ar]4s^{2}3d^{10}4p^4 selenium 78.96	35 **Br** [Ar]4s^{2}3d^{10}4p^5 bromine 79.90	36 **Kr** [Ar]4s^{2}3d^{10}4p^6 krypton 83.79
37 **Rb** [Kr]5s^1 rubidium 85.47	38 **Sr** [Kr]5s^2 strontium 87.62	39 **Y** [Kr]5s^{2}4d^1 yttrium 88.91	40 **Zr** [Kr]5s^{2}4d^2 zirconium 91.22	41 **Nb** [Kr]5s^{1}4d^4 niobium 92.91	42 **Mo** [Kr]5s^{1}4d^5 molybdenum 95.94	43 **Tc** [Kr]5s^{2}4d^5 technetium (98)	44 **Ru** [Kr]5s^{1}4d^7 ruthenium 101.1	45 **Rh** [Kr]5s^{1}4d^8 rhodium 102.9	46 **Pd** [Kr]4d^{10} palladium 106.4	47 **Ag** [Kr]5s^{1}4d^{10} silver 107.9	48 **Cd** [Kr]5s^{2}4d^{10} cadmium 112.4	49 **In** [Kr]5s^{2}4d^{10}5p^1 indium 114.8	50 **Sn** [Kr]5s^{2}4d^{10}5p^2 tin 118.7	51 **Sb** [Kr]5s^{2}4d^{10}5p^3 antimony 121.8	52 **Te** [Kr]5s^{2}4d^{10}5p^4 tellurium 127.6	53 **I** [Kr]5s^{2}4d^{10}5p^5 iodine 126.9	54 **Xe** [Kr]5s^{2}4d^{10}5p^6 xenon 131.3
55 **Cs** [Xe]6s^1 cesium 132.9	56 **Ba** [Xe]6s^2 barium 137.3	*	72 **Hf** [Xe]6s^{2}4f^{14}5d^2 hafnium 178.5	73 **Ta** [Xe]6s^{2}4f^{14}5d^3 tantalum 180.9	74 **W** [Xe]6s^{2}4f^{14}5d^4 tungsten 183.9	75 **Re** [Xe]6s^{2}4f^{14}5d^5 rhenium 186.2	76 **Os** [Xe]6s^{2}4f^{14}5d^6 osmium 190.2	77 **Ir** [Xe]6s^{2}4f^{14}5d^7 iridium 192.2	78 **Pt** [Xe]6s^{1}4f^{14}5d^9 platinum 195.1	79 **Au** [Xe]6s^{1}4f^{14}5d^{10} gold 197.0	80 **Hg** [Xe]6s^{2}4f^{14}5d^{10} mercury 200.5	81 **Tl** [Xe]6s^{2}4f^{14}5d^{10}6p^1 thallium 204.4	82 **Pb** [Xe]6s^{2}4f^{14}5d^{10}6p^2 lead 207.2	83 **Bi** [Xe]6s^{2}4f^{14}5d^{10}6p^3 bismuth 209.0	84 **Po** [Xe]6s^{2}4f^{14}5d^{10}6p^4 polonium (209)	85 **At** [Xe]6s^{2}4f^{14}5d^{10}6p^5 astatine (210)	86 **Rn** [Xe]6s^{2}4f^{14}5d^{10}6p^6 radon (222)
87 **Fr** [Rn]7s^1 francium (223)	88 **Ra** [Rn]7s^2 radium (226)	**	104 **Rf** [Rn]7s^{2}5f^{14}6d^2 rutherfordium (261)	105 **Db** [Rn]7s^{2}5f^{14}6d^3 dubnium (262)	106 **Sg** [Rn]7s^{2}5f^{14}6d^4 seaborgium (266)	107 **Bh** [Rn]7s^{2}5f^{14}6d^5 bohrium (264)	108 **Hs** [Rn]7s^{2}5f^{14}6d^6 hassium (277)	109 **Mt** [Rn]7s^{2}5f^{14}6d^7 meitnerium (268)	110 **Ds** [Rn]7s^{2}5f^{14}6d^8 darmstadtium (271)	111 **Rg** roentgenium (272)	112 **Cn** copernicium (277)	113 **Uut** (?)	114 **Uuq** (285)	115 **Uup** (?)	116 **Uuh** (289)	117 **Uus** (?)	118 **Uuo** (?)

Lanthanide Series*

57 **La** [Xe]6s^{2}5d^1 lanthanum 138.9	58 **Ce** [Xe]6s^{2}4f^{1}5d^1 cerium 140.1	59 **Pr** [Xe]6s^{2}4f^3 praseodymium 140.9	60 **Nd** [Xe]6s^{2}4f^4 neodymium 144.2	61 **Pm** [Xe]6s^{2}4f^5 promethium (145)	62 **Sm** [Xe]6s^{2}4f^6 samarium 150.4	63 **Eu** [Xe]6s^{2}4f^7 europium 152.0	64 **Gd** [Xe]6s^{2}4f^{7}5d^1 gadolinium 157.2	65 **Tb** [Xe]6s^{2}4f^9 terbium 158.9	66 **Dy** [Xe]6s^{2}4f^{10} dysprosium 162.5	67 **Ho** [Xe]6s^{2}4f^{11} holmium 164.9	68 **Er** [Xe]6s^{2}4f^{12} erbium 167.3	69 **Tm** [Xe]6s^{2}4f^{13} thulium 168.9	70 **Yb** [Xe]6s^{2}4f^{14} ytterbium 173.0	71 **Lu** [Xe]6s^{2}4f^{14}5d^1 lutetium 175.0

Actinide Series**

89 **Ac** [Rn]7s^{2}6d^1 actinium (227)	90 **Th** [Rn]7s^{2}6d^2 thorium 232.0	91 **Pa** [Rn]7s^{2}5f^{2}6d^1 protactinium 231	92 **U** [Rn]7s^{2}5f^{3}6d^1 uranium 238	93 **Np** [Rn]7s^{2}5f^{4}6d^1 neptunium (237)	94 **Pu** [Rn]7s^{2}5f^6 plutonium (244)	95 **Am** [Rn]7s^{2}5f^7 americium (243)	96 **Cm** [Rn]7s^{2}5f^{7}6d^1 curium (247)	97 **Bk** [Rn]7s^{2}5f^9 berkelium (247)	98 **Cf** [Rn]7s^{2}5f^{10} californium (251)	99 **Es** [Rn]7s^{2}5f^{11} einsteinium (252)	100 **Fm** [Rn]7s^{2}5f^{12} fermium (257)	101 **Md** [Rn]7s^{2}5f^{13} mendelevium (258)	102 **No** [Rn]7s^{2}5f^{14} nobelium (259)	103 **Lr** [Rn]7s^{2}5f^{14}6d^1 lawrencium (262)

Los Alamos National Laboratory Chemistry Division

© Copyright 2011 Los Alamos National Security, LLC. All Rights reserved. The public may copy and use this information without charge, provided that this Notice and any statement of authorship are reproduced on all copies. Neither the Government nor LANS makes any warranty, express or implied, or assumes any liability or responsibility for the use of this information.

Los Alamos
NATIONAL LABORATORY

CHAPTER 6

Names and Formulas of Chemical Compounds

In This Chapter

➤ Ionic and molecular compounds

➤ Names and formulas of ions

➤ Formulas of ionic compounds

➤ Names and formulas of molecular compounds

It's important that scientists communicate with each other. Communication requires that a standard vocabulary be adopted. In chemistry, part of that vocabulary is chemical nomenclature—the names and formulas of ions and compounds. In this chapter you will learn some of the rules chemists use to name substances and to write their formulas. Please realize that there are over 60 million known chemical compounds. There are always exceptions to the rules given in this chapter; however, they are not ones with which you need to be concerned.

Ionic and Molecular Compounds

There are three classifications of elements: metals, nonmetals, and metalloids. Metals are shown on the left side of the periodic table, nonmetals on the right, and metalloids between the two. Metalloids are elements that have properties intermediate between those of metals and nonmetals.

Chemical compounds can be classified into two categories: ionic and molecular. Ionic compounds contain ions, very often a metal ion with a positive charge and a nonmetal ion with a negative charge. The bond holding the two ions together is called the ionic bond.

figure 6.1 – Periodic Table

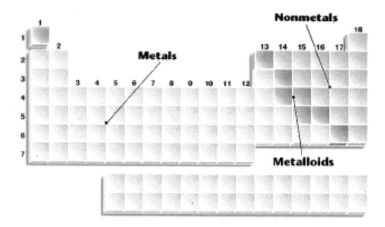

Chem Vocab

An ionic compound consists of positive and negative ions, not neutral atoms.

A molecular compound consists of neutral atoms, not ions.

Opposite charges attract each other. The force holding the ions together is the electrostatic attraction between the two opposite charges.

In molecular compounds, the atoms are all typically nonmetals, or metalloids, and are present as neutral atoms. The bonds holding the atoms together are called covalent bonds and are formed by two atoms each contributing one or more electrons to the bond.

Positive Ions

Ions are said to be simple, or monatomic, if they contain only a single atom. If they contain multiple atoms, they are called complex, or polyatomic, ions. Almost all common simple positive ions consist of a metal, a generalization that gives us our first rule about names and formulas of chemical substances:

Rule 1–Single metal atoms only form positive ions, never negative ions.

The charges on several families of metal ions can be predicted from the columns occupied by those families in the periodic table. Metals in columns IA (the alkali metals), IIA (the alkaline earths), IIIA, and the upper part of IIIB (aluminum and gallium) only form one possible ion, which gives us our second rule:

Important Point

There are two ways to number the eighteen columns in the periodic table:

➤ Using Arabic numbers, they can be numbered 1 to 18;

➤ Using Roman numerals, they can be numbered I to VIII. Hydrogen heads column IA, beryllium IIA, scandium IIIA, and so on until iron, cobalt, and nickel head a three-column block labeled VIIIA. Then the numbering starts again with copper heading column IB, zinc IIB, and so on until you reach helium, which heads column VIIIB.

Roman numerals are used in this book because they correlate with important concepts like the charges on common ions.

Rule 2–Alkali metals only form +1 ions; alkaline earths only form +2 ions; scandium, yttrium, lanthanum, actinium, aluminum, and gallium only form +3 ions.

Boron, indium, and thallium are group IIIB elements, but they have been omitted because boron is a metalloid and forms a large number of molecular compounds; indium and thallium have both +1 and +3 ions.

Naming these ions is easy since the names of the ions are just the names of the elements themselves. Thus, Na^+ is the sodium ion, Mg^{2+} is the magnesium ion, Al^{3+} is the aluminum ion, and so on, as shown in the following list:

+1		**+2**		**+3**	
lithium	Li^+	beryllium	Be^{2+}	scandium	Sc^{3+}
sodium	Na^+	magnesium	Mg^{2+}	yttrium	Y^{3+}
potassium	K^+	calcium	Ca^{2+}	lanthanum	La^{3+}
rubidium	Rb^+	strontium	Sr^{2+}	actinium	Ac^{3+}
cesium	Cs^+	barium	Ba^{2+}	aluminum	Al^{3+}
francium	Fr^+	radium	Ra^{2+}	gallium	Ga^{3+}

Transition and Post-Transition Metals

Transition metals are found in the middle block of the periodic table; post-transition metals follow the transition metals. These metals often exhibit more than one simple ion. For example, iron can be Fe^{2+} or Fe^{3+}, copper can be Cu^+ or Cu^{2+}, and tin can be Sn^{2+} or Sn^{4+}. The following list shows the charges on the most common ions of these elements:

chromium	Cr^{2+}, Cr^{3+}	gold	Au^+, Au^{3+}
manganese	Mn^{2+}, Mn^{3+}	cadmium	Cd^{2+}
iron	Fe^{2+}, Fe^{3+}	mercury	Hg_2^{2+}, Hg^{2+}
cobalt	Co^{2+}	indium	In^+, In^{3+}
nickel	Ni^{2+}	thallium	Tl^+, Tl^{3+}
copper	Cu^+, Cu^{2+}	tin	Sn^{2+}. Sn^{4+}
zinc	Zn^{2+}	lead	Pb^{2+}, Pb^{4+}
silver	Ag^+	bismuth	Bi^{3+}

The Suffixes ous and ic

There are two systems of nomenclature for simple metal ions. One system for naming these ions uses the suffixes *ous* and ic:

Rule 3–When a metallic element can exhibit ions with different charges, the ion with the lower charge gets the suffix *ous* and the ion with the higher charge gets the suffix *ic*.

The following list gives examples that use the *ous-ic* suffixes:

Cr^{2+}	chromous	Au^+	aurous
Cr^{3+}	chromic	Au^{3+}	auric
Mn^{2+}	manganous	Hg_2^{2+}	mercurous
Mn^{3+}	manganic	Hg^{2+}	mercuric
Fe^{2+}	ferrous	Sn^{2+}	stannous
Fe^{3+}	ferric	Sn^{4+}	stannic
Cu^+	cuprous	Pb^{2+}	plumbous
Cu^{2+}	cupric	Pb^{4+}	plumbic

The two mercury ions may appear puzzling since they both have a +2 charge. Notice that the mercurous ion has two atoms; each atom carries a +1 charge, which gives a total charge of +2. The charge of +1 per atom is less than the charge of +2 on the single atom in Hg^{2+}, so Hg_2^{2+} gets to be the mercurous ion and Hg^{2+} the mercuric ion.

The Stock System

An alternative to the *ous-ic* system of nomenclature is the Stock system, developed by the German chemist Alfred Stock (1876–1946) in the early twentieth century. The Stock system uses Roman numerals to indicate the charges on the metal ions. Thus, *I* indicates that the charge is +1, *II* that the charge is +2, *III* that the charge is +3, and *IV* that the charge is +4. (That is as high as charges get on metal ions.)

Since we are not using the *ous-ic* suffixes with the Stock system, the names of the ions are just the names of the elements themselves plus a Roman numeral:

Rule 4–In the Stock system the charges on metal atoms are indicated by Roman numerals written in parentheses after the name of the metal.

Important Point

Do not use the *ous* and *ic* suffixes and Roman numerals together; use one or the other, but not both at the same time.

Here are the names using the Stock system for the ions shown in the previous list:

Cr^{2+}	chromium (II)	Au^+	gold (I)
Cr^{3+}	chromium (III)	Au^{3+}	gold (II)
Mn^{2+}	manganese (II)	Hg_2^{2+}	mercury (I)
Mn^{3+}	manganese (III)	Hg^{2+}	mercury (II)
Fe^{2+}	iron (II)	Sn^{2+}	tin (II)
Fe^{3+}	iron (III)	Sn^{4+}	tin (IV)
Cu^+	copper (I)	Pb^{2+}	lead (II)
Cu^{2+}	copper (II)	Pb^{4+}	lead (IV)

An advantage of the Stock system is that you don't have to remember the Latin names of elements. However, you do have to understand Roman numerals.

Positive Ions that Do Not Contain a Metal

Among inorganic compounds there are two common examples of positively charged ions that do *not* contain metals:

➤ Hydrogen (H^+)

➤ Ammonium (NH_4^+)

The NH_4^+ ion also has the distinction of being our only example of a polyatomic positive ion besides Hg_2^{2+}. The NH_4^+ ion is about the same size as the potassium ion (K^+) and forms compounds very similar to potassium compounds. In fact, the ammonium ion is sometimes referred to as a *pseudoalkali metal ion*.

Chem Vocab

To say that an ion is a pseudoalkali metal ion is to say that it behaves like an alkali metal ion, but is not actually one.

Negative Ions

Simple negative ions—ions that contain only one atom—are formed only by nonmetals or metalloids. Several nonmetals also form complex negative ions—ions that contain more than one atom. Metals can be part of complex negative ions, but the metal atoms in the ions do not themselves have charges on them; the only charge is on the entire ion.

Important Point

There is an important exception to rule 5; hydrogen can form both an H⁺ ion and an H⁻ ion, which is called the hydride ion.

Simple Negative Ions

All simple negative ions consist of nonmetal atoms, a generalization that brings us to our fifth rule about ions:

Rule 5–Single nonmetal atoms only form negative ions, never positive ions.

Charges on simple nonmetal ions can be predicted from the columns of the periodic table in which they are found as shown in the next rule:

Rule 6–Elements located one column to the left of the noble gases (the halogens) have a charge of –1. Elements two columns to the left of the noble gases have a charge of –2. Elements three columns to the left have a charge of –3.

The names of simple nonmetal ions are obtained by dropping the suffixes from the names of the elements and substituting an *ide* suffix, as given in the next rule:

Rule 7–Names of simple nonmetal ions have a suffix of *ide*.

The following list shows the names of the most common ions that use the *ide* nomenclature:

Charge:	-1		-2		-3	
	F^-	fluoride	O^{2-}	oxide	N^{3-}	nitride
	Cl^-	chloride	S^{2-}	sulfide	P^{3-}	phosphide
	Br^-	bromide				
	I^-	iodide				

There also are two negative ions that have the *ide* suffix but that are not simple ions because they contain more than one atom: hydroxide (OH^-) and cyanide (CN^-).

Historical Note

Many chemical substances were given names before anyone knew what their compositions were. When a substance's name does not seem to be following the rules, it just means it was named before the modern rules of chemical nomenclature were developed. The older name is often retained because that's what people are in the habit of using.

Complex Negative Ions

There are numerous negative ions that contain more than one atom. These are called complex or polyatomic *ions.* If the ion contains oxygen, another name is oxyanion.

Here is a list of common oxyanions:

CO_3^{2-}	carbonate	CrO_4^{2-}	chromate
NO_3^-	nitrate	$Cr_2O_7^{2-}$	dichromate
SO_4^{2-}	sulfate	MnO_4^-	permanganate
PO_4^{3-}	phosphate	ClO^-	hypochlorite

The Suffixes ite *and* ate

There are oxyanions that have very similar formulas but which differ in the number of oxygen atoms. The suffix *ate* is generally used when the maximum possible number of oxygen atoms are bonded to the non-oxygen atom. The suffix *ite* is used to name the ion containing fewer oxygen atoms. This usage is explained in our next rule:

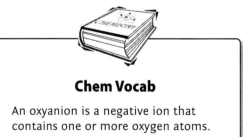

Chem Vocab

An oxyanion is a negative ion that contains one or more oxygen atoms.

Rule 8–When the oxyanions of an element can contain different numbers of oxygen atoms, the ion with fewer oxygen atoms gets the suffix *ite*, and the ion with the greater number of oxygen atoms gets the suffix *ate*.

The following list shows the names of common ions that use the *ite-ate* nomenclature:

NO_2^-	nitrite	SO_3^{2-}	sulfite
NO_3^-	nitrate	SO_4^{2-}	sulfate

Phosphate Ions

Phosphorus forms several ions. The three most common ones are phosphate (PO_4^{3-}), monohydrogen phosphate (HPO_4^{2-}), and dihydrogen phosphate ($H_2PO_4^-$).

Halogen Ions

In the case of the halogens, fluorine only forms one ion—F^-. Chlorine, bromine, and iodine, however, have several ions. The following list shows the ions for chlorine, bromine, and iodine:

ClO_4^-	perchlorate	BrO_4^-	perbromate	IO_4^-	periodate
ClO_3^-	chlorate	BrO_3^-	bromate	IO_3^-	iodate
ClO_2^-	chlorite	BrO_2^-	bromite	IO_2^-	iodite
ClO^-	hypochlorite	BrO^-	hypobromite	IO^-	hypoiodite

Chem Vocab

Chemists' use of the prefix *hypo* is similar to the use of *hypo* in the word *hypodermic*. A hypodermic needle is a needle that goes under the skin. In the list of ions, hypochlorite appears under chlorite; hypochlorite ions have fewer oxygen atoms than chlorite ions do.

Notice the prefixes *per*, from "hyper," meaning "above," and *hypo*, meaning "below" or "beneath." There are other ions that have the prefix *per*, but this is probably the only use in chemistry of *hypo*.

Formulas of Ionic Compounds

When writing formulas of ionic compounds, the positive ion is always written first and then the negative ion. Subscripts are used to indicate the numbers of ions that are in the compound.

The important thing to remember is that all compounds must be electrically neutral. Ionic compounds form by combining positive and negative ions. The number of each kind of ion must be such that the sum of the positive charges balances the sum of the negative charges. For example, when Na^+ and Cl^- form sodium chloride, only one of each ion is required because $+1 - 1 = 0$, and the formula is NaCl.

Important Point

In compounds the sum of the positive charges has to equal the sum of the negative charges.

On the other hand, when Ca^{2+} and Cl^- form calcium chloride, two chloride ions are required so that $+2 + (2)(-1) = 0$, and the formula is $CaCl_2$. If one of the ions contains more than one atom, then the formula for that ion is enclosed in parentheses, as in calcium nitrate, $Ca(NO_3)_2$.

Using the Stock system, Cu_2O is copper (I) oxide. Remember that the Roman numeral *I* indicates the charge on the copper ion, not the number of copper atoms in the formula.

The following list gives examples of the way ions combine to make electrically neutral compounds:

Ions	Formula	Names
Fe^{2+}, O^{2-}	FeO	Ferrous oxide, or iron (II) oxide
Fe^{3+}, O^{2-}	Fe_2O_3	Ferric oxide, or ion (III) oxide
Cu^{2+}, NO_3^-	$Cu(NO_3)_2$	Cupric nitrate, or copper (II) nitrate
Na^+, CO_3^{2-}	Na_2CO_3	Sodium carbonate, or sodium (I) carbonate.
Al^{3+}, SO_4^{2-}	$Al_2(SO_4)_3$	Aluminum sulfate, or aluminum (III) sulfate
Pb^{2+}, Cl^-	$PbCl_2$	Plumbous chloride, or lead (II) chloride
Pb^{2+}, Cl^-	$PbCl_4$	Plumbic chloride, or lead (IV) chloride
Ni^{2+}, PO_4^{3-}	$Ni_3(PO_4)_2$	Nickelous phosphate, or nickel (II) phosphate
Ba^{2+}, OH^-	$Ba(OH)_2$	Barium hydroxide, or barium (II) hydroxide
K^+, CN^-	KCN	Potassium cyanide, or potassium (I) cyanide
K^+, $Cr_2O_7^{2-}$	$K_2Cr_2O_7$	Potassium dichromate, or potassium (I) dichromate
NH_4^+, CO_3^{2-}	$(NH_4)_2CO_3$	Ammonium carbonate

Naming Molecular Compounds

In general, we recognize ionic compounds because they are likely to contain a metal and a nonmetal. In other words, they contain elements from the left hand side of the periodic table and the right hand side.

Molecular compounds, however, do not contain ions. Therefore, we usually think of molecular compounds as not containing metals (although we'll qualify that statement later). They usually contain only nonmetals, or a nonmetal and a metalloid. Therefore, molecular compounds most commonly only contain elements from the right side of the periodic table.

With molecular compounds, we *never* (and I really do mean *never*) use the *ous–ic* or *ite–ate* suffixes, or the Roman numerals. Instead, we use the following Greek prefixes to indicate numbers of atoms:

➤ Mono: 1

➤ Di: 2

➤ Tri: 3

➤ Tetra: 4

➤ Penta: 5

➤ Hexa: 6

➤ Hepta: 7

➤ Octa: 8

➤ Nona: 9

➤ Deca: 10

We could continue with numbers greater than ten (depending on how much Greek we know), but these are enough for our purposes.

The prefixes tri, and penta through deca, should look familiar from high school geometry, where they were used to indicate the number of sides on polygons. For example, a pentagon is a five-sided figure, a hexagon is a six-sided figure, and an octagon is an eight-sided figure. On the other hand, be careful to remember the prefixes di and tetra for the numbers 2 and 4, respectively. Sometimes people make the mistake of using bi for 2, as in bicycle, and quad for 4, as in quadrangle. Bi and quad, however, are Latin prefixes, and this system of nomenclature uses only Greek prefixes. Also, if you were thinking sept for the number 7, note that sept is Latin. Chemists do, in fact, sometimes use Latin prefixes, but in the contexts of other kinds of compounds.

The following list gives examples of the nomenclature that is used with molecular compounds. Notice that when mono and oxide are used together, one *o* is dropped. We say *monoxide*, not *monooxide*. We do the same with penta, hexa, hepta, octa, nona, and deca, dropping the *a*.

- ➤ CO Carbon monoxide
- ➤ CO_2 Carbon dioxide
- ➤ SO_2 Sulfur dioxide
- ➤ SO_3 Sulfur trioxide
- ➤ N_2O Dinitrogen oxide
                    Common name: Nitrous oxide, or laughing gas
- ➤ NO Nitrogen monoxide
                    Common name: Nitric oxide
- ➤ NO_2 Nitrogen dioxide
- ➤ N_2O_4 Dinitrogen tetroxide
- ➤ N_2O_5 Dinitrogen pentoxide
- ➤ P_2O_5 Diphosphorus pentoxide
- ➤ P_4O_{10} Tetraphosphorus decoxide
- ➤ NH_3 Nitrogen trihydride
                    Common name: ammonia
- ➤ H_2O Dihydrogen oxide
                    Common name: water!
- ➤ BH_3 Boron trihydride
                    Common name: borane
- ➤ PH_3 Phosphorus trihydride
                    Common name: phosphine
- ➤ SiO_2 Silicon dioxide
- ➤ N_2H_4 Dinitrogen tetrahydride
                    common name: hydrazine

A Question About +4 Ions

In compounds like MnO_2 and PbO_2, manganese and lead appear to be present as +4 ions. However, +4 really refers to the oxidation state of the metal (see Chapter 10), not the charge on the ion. A hard line does not exist in nature separating covalent and ionic bonding. Carbon dioxide (CO_2) clearly exhibits covalent bonding and sodium chloride (NaCl) clearly exhibits ionic bonding. Many substances, however, lie in more of a gray area. MnO_2 and PbO_2 are examples of substances that really exhibit covalent bonding, not ionic bonding. Therefore, they are often named using the rules for molecular substances and are called manganese dioxide and lead dioxide.

Using the Stock system, MnO_2 is still manganese (IV) oxide and PbO_2 is lead (IV) oxide. These two examples illustrate one of the advantages of using the Stock system when naming compounds containing metals and nonmetals—the Stock system does not rely on knowing whether the substance exhibits covalent or molecular bonding.

Trisodium Phosphate

In the paint department of probably any hardware store, you can find boxes of trisodium phosphate (Na_3PO_4) on the shelf. Substances that contain phosphate are good cleaners, and trisodium phosphate (often labeled as TSP) is also somewhat abrasive. Trisodium phosphate

Important Point

When naming molecular compounds we never use the *ous* and *ic* suffixes, nor do we use Roman numerals. When naming ionic compounds we never use the prefixes mono, di, tri, etc; the names of the ions tell us what their charges are and therefore how many of each ion there are.

is often used to clean previously painted walls prior to repainting them. According to everything that has been said so far, however, trisodium phosphate is not a correct chemical name because it is an ionic compound, and we do not use prefixes with ionic compounds. Its correct chemical name is simply "sodium phosphate." One cannot expect, however, to see commercial products always conforming to the rules of strict chemical nomenclature. In this case the manufacturer is simply trying to distinguish Na_3PO_4 from Na_2HPO_4 (sodium monohydrogen phosphate) and NaH_2PO_4 (sodium dihydrogen phosphate).

Important Point

In the Stock system of nomenclature, the Roman numeral never indicates the number of metal atoms in the formula, except maybe by coincidence. The Roman numeral always indicates the charge on the metal ion.

Conclusion

The examples in this chapter have been of inorganic compounds. Organic compounds—substances that contain carbon and hydrogen—use their own system of nomenclature and will be discussed in Chapter 30.

PART TWO

The Transformations of Matter

CHAPTER 7

Chemical Reactions

In This Chapter

➤ Conservation of mass

➤ Equations for chemical reactions

➤ Chemical reactions

➤ Stoichiometry

There are two kinds of chemical reactions: so-called ordinary chemical reactions and nuclear reactions. In ordinary chemical reactions there are no changes to the nuclei of the atoms. The only interaction between the atoms is among the atoms' electrons. In nuclear reactions the electrons do not matter. What matters are changes that take place in the atoms' nuclei. This chapter is about ordinary chemical reactions; nuclear reactions are discussed in Chapters 13 and 14.

Chem Vocab

An ordinary chemical reaction is a transformation that does not change the identities of the atoms undergoing reaction.

A nuclear reaction is a transformation that often changes the identities of the atoms undergoing reaction.

The Law of Conservation of Mass

In the latter part of the eighteenth century, the French chemist Antoine Lavoisier rose to distinction among chemists of his day due to the very careful quantitative measurements he made of substances as they were undergoing chemical reactions.

In a chemical reaction we start with elements or compounds that we call the reactants. What takes place in a chemical reaction is that bonds between atoms in the compounds that make up the reactants are broken and the atoms are rearranged to form new compounds we call the products of the reaction.

Historical Note

The law of conservation of mass played an important role in the determination of formulas of chemical compounds. A compound can be decomposed into its individual elements. Each element can then be weighed. From the ratios of the masses of the elements in the compound, chemists can work backwards and figure out what the ratios of the atoms are in the compound.

What Lavoisier discovered is that there is no discernible change in mass during a chemical reaction. This means that if we carefully weigh all of the substances that are going to react with each other, and then carefully collect and weigh all of the products that form, within the limits of very precise analytical balances, there is neither an increase nor a decrease in mass. Mass is conserved. Lavoisier stated this as the law of conservation of mass.

Important Point

Today, because of Albert Einstein's theory of relativity, we know that mass is *not* strictly conserved. Einstein showed that mass and energy are equivalent, as stated in his famous equation $E = mc^2$, where E = energy, m = change in mass, and c = speed of light. Changes in mass are appreciable in nuclear reactions. However, in ordinary chemical reactions the changes are so miniscule that they can be ignored.

Writing Equations for Chemical Reactions

The general form for writing an equation for a chemical reaction is to write the formulas for the reactants on the left-hand side and the formulas for the products on the right-hand side, with an arrow pointing from the reactants to the products:

$$\text{Reactants} \rightarrow \text{Products.}$$

For example, in the combustion of propane gas, C_3H_8, propane combines with oxygen from the atmosphere to give carbon dioxide and water (and heat, which we will ignore for now, but to which we will return in a later chapter). The unbalanced equation for the reaction is

$$C_3H_8 \text{ (g)} + O_2 \text{ (g)} \rightarrow CO_2 \text{ (g)} + H_2O \text{ (g),}$$

where *g* means the substance is a gas. (There are other symbols we use, too: *l* = liquid, *s* = solid, *aq* = aqueous solution.)

The reason this equation is unbalanced is because there are unequal numbers of carbon, hydrogen, and oxygen atoms on each side of the equation. It looks like atoms of some elements have been turned into atoms of other elements! But that only happens in a nuclear reaction, and this is not a nuclear reaction. Being unbalanced also is a violation of the law of conservation of mass. Right now, if we added together the masses of the atoms as shown on the left, and then added together the masses of the atoms as shown on the right, we would get different sums.

To correct our equation we balance it, that is, we place numerical coefficients in front of one or more of the substances shown to give us equal numbers of each kind of atom. The balanced equation looks like this:

$$C_3H_8 \text{ (g)} + 5\,O_2 \text{ (g)} \rightarrow 3\,CO_2 \text{ (g)} + 4\,H_2O \text{ (g).}$$

Let us check to ensure that it is in fact balanced:

Left side	Right side
3 carbon atoms	3 carbon atoms
8 hydrogen atoms	8 (=4x2) hydrogen atoms
10 (=5x2) oxygen atoms	10 (=[3x2]+[4x1]) oxygen atoms.

To be considered correctly written, all equations for chemical reactions must be balanced.

Examples of Chemical Reactions

In this book we will consider a variety of reactions that take place in the gaseous state, in liquids other than water (nonaqueous media), and in aqueous solutions. Let us consider a few examples before we move on to increase our understanding of chemical reactions and how we interpret equations that represent those reactions.

Reactions in the Gas Phase

Most of the time you might not be conscious of it, but chemical reactions are taking place in the atmosphere all the time. If the sky above you is not blue, though, or if there is an odor to the air around you, then you know that chemical reactions have taken place.

For example, ground-level air pollution usually begins with the exhaust gases being emitted into the atmosphere from motor vehicles. The exhaust gases consist mostly of carbon dioxide (CO_2) and water (H_2O), but they may contain some nitric oxide (NO). Catalytic converters are supposed to remove NO, and they mostly do, but air pollution does take place.

Nitric oxide itself is a colorless gas, but in the atmosphere it is converted to nitrogen dioxide (NO_2) by the following reaction:

$$2 \, NO \, (g) + O_2 \, (g) \rightarrow 2 \, NO_2 \, (g).$$

If you have ever been in an urban environment in warm weather, you probably saw NO_2, a reddish-brown gas that lends a brownish haze to the sky if there is enough of it. From an airplane or a hilltop, a layer of brown-colored air is particularly conspicuous. NO_2 is also toxic, but it is not the real culprit in an air pollution episode.

The real culprit is ozone (O_3). Ozone is a byproduct of the action of sunlight on NO_2, as shown in the following reactions:

$$NO_2 \, (g) + sunlight \rightarrow NO \, (g) + O \, (g);$$

$$O_2 \, (g) + O \, (g) \rightarrow O_3 \, (g).$$

Most people refer to a brownish-color sky as smog, but the more correct term is photochemical haze— *photochemical* because it is driven by sunlight, and *haze* because visibility is reduced.

Numerous other chemical reactions take place in the gas-phase, but we will return to them later.

Reactions in Aqueous Solution

Much of Earth is covered with water. Water is our most common liquid and is a universal solvent, dissolving many more substances than most other liquids do. Many chemical reactions in water take place between ions. Two common kinds of reactions that take place in water are precipitation reactions and acid-base reactions.

Precipitation Reactions

A precipitation reaction is simply a chemical reaction in which a solid (called a precipitate) forms when two solutions are mixed together. Usually what happens is that a positive ion and a negative ion combine to form an electrically neutral compound that is insoluble in water. The scale that forms inside a tea kettle is an example. Tap water typically contains calcium ions (Ca^{2+}) and carbonate ions (CO_3^{2-}). Calcium carbonate ($CaCO_3$) precipitates over time, so it builds up inside tea kettles (and coffee pots, hot water heaters, and car radiators) as shown by the following reaction:

$$Ca^{2+} (aq) + CO_3^{2-} (aq) \rightarrow CaCO_3 (s).$$

Calcium carbonate is a white solid. In nature, it occurs in the form of several minerals, including limestone, marble, and travertine. Sea shells are made of calcium carbonate.

Conservation of Charge

This reaction illustrates a second conservation law that must be taken into consideration when balancing chemical equations. (The first law was the law of conservation of mass, according to which we have to make sure that we have the same number of atoms of each element on both sides of the equation.) This law is the law of conservation of charge, which in nature is actually more fundamental than conservation of mass.

We cannot create or destroy electrical charge in a chemical reaction. Therefore, the *sum* of the charges on the right-hand side of the equation has to equal the sum of the charges on the left-hand side. The charges on the left side were +2 and –2, which add to zero. The only substance on the right side is a neutral compound, whose charge is also zero, so we see that the charges do balance. It is important to remember, though, anytime a reaction has ions in it to make sure that charge is conserved.

Important Point

The law of conservation of charge states that electrical charge is never created or destroyed. In any physical or chemical process, charge is always conserved.

Acid-base Reactions

Vinegar contains a weak acid called acetic acid ($HC_2H_3O_2$). Some drain openers contain lye, which is sodium hydroxide (NaOH) and is a strong base. If we mix the two together, the following reaction takes place:

$$HC_2H_3O_2 \ (aq) + NaOH \ (aq) \rightarrow H_2O \ (l) + NaC_2H_3O_2 \ (aq).$$

This reaction is different from the reaction that formed $CaCO_3$ because there are no solids in this reaction. The label *aq* reminds us that these substances are dissolved in water. Substances that are dissolved in water can be molecular or ionic, but are not solids. By definition, solids are insoluble in water.

Other Chemical Reactions

We will encounter a large number of chemical reactions in this book. To appreciate them, pay attention to the nature of the reactants—whether they are solids, liquids, or gases, and whether they are inorganic substances or organic substances. Look to see how the atoms in the reactants have been rearranged to form the product species. If you see the symbol *s* labeling a product, try to picture a solid that has precipitated. If you see a *g* labeling a product, try to picture gas bubbles. With experience you will also learn to visualize colors of solids or colors of solutions. If gases are noxious substances, like hydrogen sulfide (H_2S) or nitrogen dioxide (NO_2), you might also try to imagine the odors that are produced.

Stoichiometry of Chemical Reactions

The word *stoichiometry* refers to mass relationships in chemical reactions. Typically, stoichiometric calculations tell how much product can be formed from a given amount of starting materials. This section explains how we answer that kind of a question.

Molecular Weight

First, we need to know how to calculate molecular weights of compounds. Two numbers are given in the periodic table for each element. The first number is a whole number; recall from Chapter 4 that it is called the atomic number of the element. The second number is a decimal number and is called the atomic weight of the element. An element's atomic weight is the average mass number of the element's various isotopes as found here on Earth. For example, the atomic weight of mercury is 200.59.

There are two ways to interpret that number. The first is to say that a single atom of mercury has a mass of 200.59 atomic mass units (abbreviated amu, or sometimes just u.) Of course, there are no laboratory balances capable of weighing a single atom, so no balance is actually calibrated in atomic mass units. Another interpretation, however, is to say that one mole of mercury atoms has a mass of 200.59 grams. That is the interpretation of atomic weight we will use.

Chem Vocab

Stoichiometry is the method by which the quantities of reactants and products in chemical reactions are determined.

Important Point

The molecular weight of a compound is calculated by adding together the atomic weights of all the atoms that are in a molecule of the compound.

The molecular weight of a compound is just the sum of the atomic weights of the atoms found in a molecule of that compound. Therefore, if we make mercuric oxide (HgO) from mercury and oxygen, we can calculate that the mass of one mole of mercuric oxide is 200.59 + 16.00 = 216.59 grams. Its molecular weight, therefore, is 216.59 grams/mole. Similarly, if we want to know the molecular weight of calcium nitrate [$Ca(NO_3)_2$], we add together the atomic weight of calcium (40.08 g/mol), plus two times the atomic weight of nitrogen (14.01 g/mol), plus six times the atomic weight of oxygen (16.00 g/mol). The calculation looks like this:

$$40.08 + (2)(14.01) + (6)(16.00) = 164.10;$$

we say that the molecular weight of $Ca(NO_3)_2$ is 164.10 g/mol.

Calculating the Mass of a Product in a Chemical Reaction

Consider again the combustion of propane gas (C_3H_8) in an excess of oxygen gas. This is the reaction:

$$C_3H_8 \ (g) + 5\ O_2 \ (g) \rightarrow 3\ CO_2 \ (g) + 4\ H_2O \ (g).$$

Suppose we burn completely 200 g of C_3H_8. How many grams of CO_2 and H_2O would be formed? Here are the steps:

➤ We need to calculate the molecular weights of C_3H_8, CO_2, and H_2O. Doing so we get 44.11 g/mol for C_3H_8, 44.01 g/mol for CO_2, and 18.02 g/mol for H_2O.

➤ We use the coefficients in the balanced equation to find the mole ratios of the compounds involved in the reaction. We see that the ratio of CO_2 to C_3H_8 is 3 to 1, and the ratio of H_2O to C_3H_8 is 4 to 1.

➤ We solve this problem by setting up the following set of calculations:

$$200\ \text{g}\ C_3H_8 \ \times\ \frac{1\ \text{mol}\ C_3H_8}{44.11\ \text{g}\ C_3H_8}\ \times\ \frac{3\ \text{mol}\ CO_2}{1\ \text{mol}\ C_3H_8}\ \times\ \frac{44.01\ \text{g}\ CO_2}{\text{mol}\ CO_2}\ =\ 599\ \text{g}\ CO_2$$

$$200\ \text{g}\ C_3H_8 \ \times\ \frac{1\ \text{mol}\ C_3H_8}{44.11\ \text{g}\ C_3H_8}\ \times\ \frac{4\ \text{mol}\ H_2O}{1\ \text{mol}\ C_3H_8}\ \times\ \frac{18.02\ \text{g}\ H_2O}{\text{mol}\ H_2O}\ =\ 327\ \text{g}\ H_2O$$

Notice how units that are in both a numerator and a denominator cancel. The answer to our question is that the complete combustion of 200 grams of C_3H_8 results in the formation of 599 grams of CO_2 and 327 grams of H_2O.

Conclusion

Chemistry is all about chemical reactions. Try to become familiar with the formulas of more familiar substances. When reading a chemical equation, try to picture what is happening, whether it is a solid that is precipitating, a gas that is bubbling from a solution, or a brilliant change in color of a solution.

CHAPTER 8

Precipitation Reactions

In This Chapter

➤ Solubility rules
➤ Reactions in aqueous solution

Many chemical reactions take place in aqueous solution, that is, solutions in which water is the solvent. Aqueous solutions are everywhere: aquatic ecosystems, including oceans, lakes, rivers, ponds, and puddles; beverages, including dairy products, soft drinks, alcoholic beverages, and drinking water; the fluids of living organisms (which are mostly water), including blood, sweat, mucous, and urine; swimming pools, to which chemicals are added to disinfect and adjust the acidity of the water; lead storage batteries, which are filled with sulfuric acid; and laboratories, whose shelves are lined with bottled solutions.

Chem Vocab

An ecosystem consists of living organisms and the nonliving environment in which they are found.

Solubility Rules

Refer to Table 8.1. Let's start with the general rules and then work our way down to more specific ones.

Table 8.1: Solubility Rules

> ➤ All salts containing alkali metal ions, e.g., Na^+ and K^+, are soluble. All NH_4^+, $C_2H_3O_2^-$, ClO_3^-, ClO_4^-, and NO_3^- salts are soluble.

> ➤ All Cl^-, Br^-, I^-, and SCN^- salts are soluble, except those of Ag^+, Pb^{2+}, and Hg_2^{2+}. ($PbCl_2$ is soluble in hot water.)

> ➤ All F^- salts are soluble except those of the alkaline earth metals and Pb^{2+}.

> ➤ All SO_4^{2-} salts are soluble except those of Ca^{2+}, Ba^{2+}, Sr^{2+}, Hg_2^{2+}, and Pb^{2+}. (Ag^+ and Hg^{2+} sulfates are slightly soluble.)

> ➤ All O^{2-} and OH^- compounds are insoluble except those of Na^+, K^+ (and other alkali metals), NH_4^+, Ca^{2+}, Ba^{2+}, and Sr^{2+}.

> ➤ All CO_3^{2-}, PO_4^{3-}, and SO_3^{2-} compounds are insoluble except those of Na^+, K^+ (and other alkali metals), and NH_4^+.

> ➤ All S^{2-} compounds are insoluble except those of the alkali and alkaline earth metals and NH_4^+.

> ➤ All $C_2O_4^{2-}$ compounds are insoluble, except those of Na^+, K^+, and NH_4^+. (MgC_2O_4 is slightly soluble.)

> ➤ All CrO_4^{2-} compounds are insoluble except those of Na^+, K^+, NH_4^+, Mg^{2+}, and Ca^{2+}.

Rule 1 tells us that a lot of common compounds are soluble in water.

Sodium and potassium salts are everywhere in nature, in many food stuffs, and in our bodies. You know from experience that when you add table salt (NaCl) to water, the salt dissolves—at least up to a limit. If you add 1 gram of salt to a quart of water, it easily dissolves. You can add another gram and it dissolves. You can add 10 grams, and it dissolves. You can add 100 grams of salt, and it dissolves. There are limits, however. If you add 1,000 grams (1 kilogram) of salt to a quart of water, it probably will not dissolve completely.

In any discussion of solubility, unless we state otherwise, it is safe to assume that we are talking about dilute solutions. Adding 10 grams of salt to a quart of water creates a very dilute solution. Adding 1,000 grams is a very concentrated solution.

Any solid that is soluble in water at low concentrations has a limit to how much of it can dissolve. When the solution is holding all of the salt that it can, we say the solution is saturated. Any more salt that is added will not dissolve but will just settle on the bottom of the container.

There are very few nitrate minerals in nature due to their high solubility in water. Minerals such as soda niter ($NaNO_3$) are found mostly in dry environments like deserts. Saltpeter (KNO_3) tends to be found mostly on surfaces that are protected from rain, such as in caves or on sheltered cliff walls.

Chem Vocab

In a saturated solution the maximum amount of solute has dissolved that the solution can hold.

The high solubility of sodium and potassium salts explains the saltiness of sea water. We can recover substances like sodium chloride (NaCl) and potassium chloride (KCl) by evaporating sea water. Salts like NaCl are also recovered from salt mines, in which NaCl occurs as the mineral halite (rock salt). Salt deposits are just the remnants of ancient seas that have long since evaporated.

The remaining rules in Table 8.1 are stated in terms of the negative ions.

Rule 2 tells us that the majority of compounds that contain the halide ions Cl^-, Br^-, and I^- are soluble. Only a few halide compounds are insoluble, for example, AgBr, PbI_2, and Hg_2Cl_2. (Recall that Hg_2Cl_2 is mercurous chloride. Mercuric chloride—$HgCl_2$—is soluble in water.) We do not encounter the thiocyanate ion very often, but compounds like AgSCN are also insoluble. The Cl^- and SCN^- compounds are white in color, while the Br^- and I^- compounds are yellow in color.

Important Point

Substances that contain Na^+, K^+, or NO_3^- ions are very common. In dilute solutions, these substances are always soluble.

Rule 3 addresses the remaining halide ion—F^-. It states that all F^- salts are soluble except those of the alkaline earth metals and Pb^{2+}. The chemistry of chlorine, bromine, and iodine are all very similar to each other. Fluorine is different. The difference is due to the much smaller size of fluorine atoms compared to the other three (see Chapter 17).

Most substances that contain F^- are soluble. A common fluoride mineral is fluorite, CaF_2, which occurs in a variety of colors including purple, green, and yellow. Likewise, in rule 4, most substances that contain SO_4^{2-} are soluble, mostly with the exception of some of the alkaline earths and the lead ion.

Rule 4 applies to compounds that contain the sulfate ion—SO_4^{2-}.

The most common sulfate compounds are $SrSO_4$, $BaSO_4$, and $PbSO_4$, which are all white solids. Minerals that contain these compounds include celestite ($SrSO_4$), barite ($BaSO_4$), and anglesite ($PbSO_4$). Minerals that contain calcium sulfate ($CaSO_4$) are relatively common, and include gypsum and glauberite. Epsom salts contain magnesium sulfate ($MgSO_4$), which is soluble in water.

Compounds that contain the ions listed in rules 5–9 are mostly insoluble. The exceptions, as given in rule 1, are the alkali metals and the ammonium ion since all of their compounds are soluble.

Rule 5 applies to compounds that contain the oxide or hydroxide ions—O^{2-} or OH^-, respectively. Oxides and hydroxides are abundant in nature and include many of our economically important minerals like corundum and bauxite (both Al_2O_3), hematite (Fe_2O_3), rutile (TiO_2), quartz (SiO_2), cassiterite (SnO_2), brucite [$Mg(OH)_2$], and zincite (ZnO). Most of these are white, with the exception of hematite, which is black or dark reddish-brown.

Rule 6 addresses compounds that contain the carbonate ion—CO_3^{2-}. Some of the most common minerals in nature are various forms of calcium carbonate ($CaCO_3$) and include calcite, limestone, aragonite, travertine, chalk, and marble, all of which are white in color. Some limestone consists of very large formations that are now exposed to view. The famed white cliffs of Dover, England; a large layer of the Grand Canyon in Arizona; and an exposed layer of limestone in the Guadalupe Mountains of Texas, which date from the Permian Era, are all made of calcium carbonate.

Minerals made of calcium carbonate are sedimentary materials, and are likely to be significant sources of fossils. Fossils tend to be destroyed during the geological processes that form igneous and metamorphic rocks. Minerals like limestone accumulated over millions of years at the bottom of the oceans and then rose to the surface because of plate tectonics. Because they are relatively unaltered by the uplift process, the fossils they contain tend to still be intact.

Chem Vocab

Geologists classify rocks into three kinds. Igneous rocks have their origins in volcanoes and consist of material that has solidified from molten material. Sedimentary rocks are formed by compaction and hardening of minerals and animal remains, usually on the bottoms of oceans. Metamorphic rocks are new minerals formed by the effects of heat and pressure on igneous or sedimentary materials.

Other economically important carbonate containing minerals include dolomite [CaMg(CO$_3$)$_2$], smithsonite (ZnCO$_3$), both of which are white, and two important ores of copper—malachite [Cu$_2$CO$_3$(OH)$_2$], which is green, and azurite [Cu$_3$(OH)$_2$(CO$_3$)$_2$], which is dark blue.

Phosphates and sulfites are less common, although phosphates may be important sources of phosphorus in the form of fertilizer. Turquoise is a phosphate of copper, and is used as a gem stone.

Sulfur is the sixteenth most abundant element on Earth, so it is not surprising that sulfide (S^{2-}) compounds are extremely common. Rule 7 tells us that, with the exception of the alkali metals, the alkaline earths, and the ammonium ion, all sulfide compounds are insoluble in water. Many are extremely insoluble. Therefore, sulfides are among our most economically recoverable sources of important metals and include argentite (Ag$_2$S, gray or black), galena (PbS, gray or black), sphalerite (ZnS, white, yellow, red, or black), covellite (CuS, blue, purple, or black), cinnabar (HgS, bright red), realgar (AsS, orange-red), orpiment (As$_2$S$_3$, yellow), stibnite (Sb$_2$S$_3$, gray), and pyrite (FeS$_2$, yellow), also known as fool's gold.

Important Point

Most substances that contain metal ions, especially transition metal ions, are insoluble. The exceptions are compounds that contain alkali metals or nitrate ions. Sulfide ores are extremely common in nature and tend to be very insoluble.

Rules 8 and 9 apply to the oxalate and chromate ions— $C_2O_4^{2-}$ and CrO_4^{2-}, respectively—which are less common in nature than the ions in the preceding rules. Kidney stones are an example of an oxalate. Oxalic acid (H$_2$C$_2$O$_4$) occurs in certain foods like spinach and rhubarb. Calcium ions (Ca^{2+}) are an essential electrolyte in the body. Put Ca^{2+} and $C_2O_4^{2-}$ together in sufficient concentrations, and calcium oxalate (CaC$_2$O$_4$) precipitates.

As the kidneys filter waste substances from urine, it is possible for calcium oxalate to precipitate. Very tiny kidney stones usually will be excreted during urination without a person being aware of even having them. Larger stones, however, can become lodged in the ureters (the tubes that connect the kidneys to the bladder), an experience which is extremely painful. The stones need to be removed; otherwise, obstruction of the ureter can lead to a urinary tract infection. Removal is accomplished by one of several methods: drinking sufficient fluid to dilute the solution and flush the stone, disintegration of the stone using ultrasonic waves, or surgery.

Chromate-containing minerals are unusual, but an example of a fairly common chromate substance is crocoite, which is a form of lead chromate ($PbCrO_4$) and is a brilliant orange color.

Solubility also depends on temperature. Many substances become more soluble in water as the temperature increases and less soluble as the temperature decreases. With a few substances, the opposite is true. In this discussion, however, we are assuming that precipitation reactions are taking place at roughly room temperature: 68°—78°F (20°—25°C).

Precipitation Reactions

Precipitation reactions take place between positive ions and negative ions. In aqueous solutions all ions have a tendency to be hydrated, which means they are surrounded by water molecules. Water molecules are electrically neutral, but there is slightly more negative charge on the oxygen atom and slightly more positive charge on the hydrogen atoms. Positive ions are attracted to the negative end of water molecules around them and to any negative ions that are also present in the solution. Negative ions are attracted to the positive end of water molecules and to any positive ions that are also present. All of the forces of attraction are electrostatic in nature. Like charges repel each other. Opposite charges attract each other.

If a solid dissolves in water, it means that the forces of attraction between the ions in the solid and the surrounding water molecules are stronger than the forces of attraction between the ions themselves. If a solid is insoluble, it means that the electrostatic forces between the solid's positive and negative ions are stronger than the forces between the ions and the surrounding water molecules.

Chem Vocab

Chemists use the term *precipitate* to describe a solid that forms in a solution.

For example, consider what happens when aqueous solutions of magnesium sulfate ($MgSO_4$) and sodium hydroxide (NaOH) are mixed together. The result is the formation of a white precipitate. The following section explains how to find the formula for the precipitate, and what the balanced equation is for the reaction.

Total Molecular Equations

In the case of reacting $MgSO_4$ and NaOH, we can use the solubility rules to answer the first question. The ions that are present in the two solutions are Mg^{2+}, SO_4^{2-}, Na^+, and OH^-. In the original solutions, $MgSO_4$ is soluble because the attraction of Mg^{2+} and SO_4^{2-} ions to the surrounding water molecules is stronger than the attraction between Mg^{2+} and SO_4^{2-}. The same is true for Na^+ and OH^-. When the two solutions are mixed together, Mg^{2+} and OH^- feel an attraction to each other, and Na^+ and SO_4^{2-} feel an attraction. Rule 1 tells us that all sodium salts are soluble. Therefore, Na_2SO_4 is soluble; the precipitate is not Na_2SO_4. On the other hand, rule 5 tells us that $Mg(OH)_2$ is insoluble. Therefore, the precipitate must be $Mg(OH)_2$.

Using that information we can now write a balanced equation for the reaction that took place:

$$MgSO_4\ (aq) + 2\ NaOH\ (aq) \rightarrow Mg(OH)_2\ (s) + Na_2SO_4\ (aq).$$

This is an example of what chemists call a total molecular equation. A total molecular equation shows all of the reactant species as molecular species.

Total Ionic Equations

A more descriptive way of showing the reaction between $MgSO_4$ and $NaOH$ is to write soluble ionic compounds as their ions, rather than as whole molecules. That gives us the total ionic equation for the reaction:

$$Mg^{2+}\ (aq) + SO_4^{2-}\ (aq) + 2\ Na^+\ (aq) + 2\ OH^-\ (aq)$$
$$\rightarrow Mg(OH)_2\ (s) + 2\ Na^+\ (aq) + SO_4^{2-}\ (aq)$$

In the total molecular equation, the presence of ions was implicit. In the total ionic equation, the ions are shown explicitly. You can see at a glance which species are present as ions and which species are present as molecules.

Net Ionic Equations

We note in the above equation that we now have SO_4^{2-} and Na^+ ions on both sides of the arrow. In an algebraic equation, you can cancel terms that appear on both sides of an equals sign. Likewise, in a chemical equation, you can cancel chemical species that appear on both sides of an arrow. Cancelling SO_4^{2-} and Na^+ from both sides leaves us with the following equation:

$$Mg^{2+}\ (aq) + 2\ OH^-\ (aq) \rightarrow Mg(OH)_2\ (s).$$

This is what chemists call a net ionic equation. The term *net* means that ions that did not react are omitted. The net ionic equation shows only the ions that actually reacted.

Chem Vocab

Spectator ions are ions that are present in solutions but are unchanged by whatever chemical reactions are taking place in the solution.

Spectator Ions

In the case of precipitation reactions, we can identify which ions did not react using the solubility rules. We call these ions spectator ions. In a football stadium, the action is taking place on the playing field. The spectators are sitting in the stands. The spectators are present in the stadium but do not participate in the action taking place on the field. Spectator ions are like that—they are present in the solution but do not participate in the reaction that is taking place. Thus, a net ionic equation is one that leaves out any spectator ions that might be present.

Important Point

The whole idea of a net ionic equation is to show only the ions and molecules that actually react. Any other ions that are present just to give solutions net neutral charges are called spectator ions and are left out of the net ionic equation.

The ions listed in solubility rule 1 are always spectator ions in all precipitation reactions. Therefore, we would never write a net ionic equation that contains Na^+, K^+, NH_4^+, $C_2H_3O_2^-$, ClO_3^-, ClO_4^-, or NO_3^-. The negative ions listed in rules 2–9 could be spectator ions in some precipitation reactions, but they are not spectator ions in all precipitation reactions. For example, using rule 2, in the reaction between $AgNO_3$ and KCl, we would write the following net ionic equation:

$$Ag^+(aq) + Cl^-(aq) \rightarrow AgCl\ (s)$$

in which NO_3^- and K^+ have been left out because they are spectator ions, but in which Cl^- is not a spectator ion. On the other hand, in the reaction between $BaCl_2$ and Na_2SO_4, we would write this net ionic equation:

$$Ba^{2+}(aq) + SO_4^{2-}(aq) \rightarrow BaSO_4\ (s).$$

This time Cl^- is a spectator ion (along with Na^+).

Any ion not listed in solubility rule 1 can be a spectator ion some of the time and part of a precipitate at other times. Also, it is possible for a precipitation reaction to have only one spectator ion. It is also possible that a reaction has *no* spectator ions if both products are insoluble. An example is the reaction between $Ba(OH)_2$ and $MgSO_4$. Since $BaSO_4$ and $Mg(OH)_2$ are both insoluble, the net ionic equation is as follows:

$$Ba^{2+} (aq) + 2\ OH^- (aq) + Mg^{2+} (aq) + SO_4^{2-} (aq)$$
$$\rightarrow BaSO_4\ (s) + Mg(OH)_2\ (s).$$

More Examples of Net Ionic Equations

Solutions of $Bi(NO_3)_3$ and Na_2S are both colorless. If we mix the two solutions together, a black solid precipitates. Using solubility rule 7 we see that all sulfides are insoluble with only a few exceptions. The bismuth ion (Bi^{3+}) is not one of the exceptions. Since we know from rule 1 that $NaNO_3$ is soluble, we conclude that Na^+ and NO_3^- are spectator ions in this reaction and that the black precipitate is bismuth sulfide (Bi_2S_3), as shown in the following net ionic equation:

$$2\ Bi^{3+} (aq) + 3\ S^{2-} (aq) \rightarrow Bi_2S_3\ (s).$$

Notice that the laws of conservation of mass and conservation of charge are both satisfied in this equation. There are two atoms of bismuth on each side and there are three atoms of sulfur on each side. The sum of the charges on the left-hand side is $(2)(+3) + (3)(-2) = 0$. The charge on the right-hand side is also zero.

Just one more example before we move on to the next chapter. An aqueous solution of $FeCl_3$ is yellow. An aqueous solution of NaOH is colorless. If we mix the two solutions together, we get a rusty brown precipitate. We know from rule 1 that NaCl is soluble. Rule 6 tells us that $Fe(OH)_3$ is insoluble. Therefore, in this reaction Na^+ and Cl^- are spectator ions and the precipitate is $Fe(OH)_3$. This reaction is shown in the following net ionic equation:

$$Fe^{3+} (aq) + 3\ OH^- (aq) \rightarrow Fe(OH)_3\ (s).$$

Ferric hydroxide has a rusty brown color because, in fact, that is what rust is—either $Fe(OH)_3$ or Fe_2O_3.

Conclusion

Throughout the remainder of this book we will usually write net ionic equations for reactions that take place in aqueous solution. For reactions occurring in the gas phase, or between liquids other than water, we will continue to write total molecular equations.

CHAPTER 9

Acids and Bases

> ## In This Chapter
>
> ➤ Arrhenius acids and bases
> ➤ Brőnsted-Lowry acids and bases
> ➤ Strengths of acids and bases
> ➤ The pH scale.

Acids and bases are all around us. Soft drinks and vinegar are acidic. Rainwater is slightly acidic. Window cleaners and drain openers are basic. Most fluids in our bodies are slightly acidic or slightly basic, although stomach acid is very acidic. Let's find out what acids and bases are.

Chem Vocab

We describe solutions as being acidic, neutral, or basic. A neutral solution is neither acidic nor basic. Another word that means the same as *basic* is *alkaline*.

Definitions of Acid and Bases

During the years of alchemy the chemical compositions of acids and bases were unknown. Alchemists had to rely on the behavior of substances to classify them as acids, bases, or neither. Generally speaking, solutions of acids tasted sour. Solutions of bases tasted bitter and felt slippery. (Bases feel soapy because soaps are made of bases.)

Alchemists knew that litmus turns red in the presence of acids and blue in the presence of bases. Most metals were known to dissolve in common acids. (Gold was a notable exception. Its inability to dissolve was one source of some of the fascination alchemists had with gold.) Bases reacted with acids to form salts. Acidic solutions caused some common rocks to fizz—limestone is a particularly well known example.

Historical Note

The Latin word *acidus* means "sharp," or "sour," hence our expression of referring to an ill-tempered person as having an acid wit.

These descriptions of acids and bases lasted hundreds of years, but they told alchemists nothing about the elements of which acids or bases were made. Even after the discovery of hydrogen by the English chemist Henry Cavendish (1731–1810) in 1766, it still took chemists over a century to begin to define acids and bases in terms of their chemical compositions. The first to do so was the Swedish chemist Svante Arrhenius (1859–1927), who received the Nobel prize in chemistry in only the third year in which Nobel prizes were awarded (1903).

The Arrhenius Definition

Arrhenius's definition of an acid recognized the importance of the hydrogen ion, H^+. Arrhenius said that a chemical species is an acid if it can dissociate in aqueous solution and yield H^+ ions to the solution. H^+ ions were responsible for the sour taste of acids, for the effect of acids on litmus paper, and for the reaction of acids with metals and sedimentary materials like limestone. Examples of Arrhenius acids included hydrochloric acid (HCl), nitric acid (HNO_3), sulfuric acid (H_2SO_4), acetic acid ($HC_2H_3O_3$), phosphoric acid (H_3PO_4), and oxalic acid ($H_2C_2O_4$).

Arrhenius went on to say that a chemical species is a base if it can dissociate in aqueous solution and yield hydroxide ions (OH^-) to the solution. OH^- ions are responsible for the bitter taste and slippery feel of bases, the effect of basic substances on litmus paper, and the reaction of bases with acids to form salts. Examples of Arrhenius bases included NaOH, KOH, $Ba(OH)_2$, $Ca(OH)_2$, and $Mg(OH)_2$.

The Arrhenius definition of acids and bases explains neutralization reactions. In neutralization reactions, an acid reacts with a base to produce a salt plus water. For example, the following equation shows the reaction between HCl and NaOH:

$$HCl\ (aq) + NaOH\ (aq) \rightarrow Na^+\ (aq) + Cl^-\ (aq) + H_2O\ (l).$$

Arrhenius would have said that the driving force behind the tendency for acids and bases to react with each other is the strong affinity H^+ and OH^- ions have for each other. Given the opportunity, H^+ and OH^- ions will always react to form water molecules.

Although Arrhenius is justly credited with an attempt to define acids and bases in terms of their chemical compositions, he fell short of explaining the behavior of substances that turn litmus red but which do not contain H^+ ions in their formulas, or substances that turn litmus blue but which do not contain OH^- ions in their formulas. It was a good start, but a more complete theory of acids and bases was introduced by chemists in Denmark and England.

The Brönsted-Lowry Definition of Acids and Bases

The definition of acids and bases we will use in this book is called the Brönsted-Lowry definition, named for Johannes Brönsted (1879–1947), a Danish chemist, and Thomas Lowry (1874–1936), an English chemist, who put forth definitions of acids and bases independently of each other in 1923.

Chem Vocab

An Arrhenius acid contains one or more hydrogen ions in its formula.

An Arrhenius base contains one or more hydroxide ions in its formula.

The reaction between an Arrhenius acid and an Arrhenius base results in the formation of a salt plus water.

In their definition of an acid, the important chemical species is the hydrogen ion (H^+), often written as the hydronium ion (H_3O^+). A hydronium ion is a water molecule, H_2O, with an H^+ ion attached to the oxygen atom. The reason we often write the H^+ ion as H_3O^+ is to acknowledge the fact that most neutral hydrogen atoms consist only of a proton and an electron. To form the H^+ ion, the electron is removed, leaving only the proton. Protons are incredibly tiny centers of positive charge that are not going to exist in water as independent chemical species. Rather, they will always attach themselves to water molecules. With bases, the important chemical species is the hydroxide ion (OH^-).

According to Brönsted and Lowry, a molecule or ion is an acid if, when the species is added to water, the concentration of H^+ (or equivalently, H_3O^+) increases. Consider, for example, what happens when hydrochloric acid is added to water:

$$HCl\ (aq) + H_2O\ (l) \rightarrow Cl^-\ (aq) + H_3O^+\ (aq).$$

The formation of H_3O^+ ions makes the solution acidic. Another example is the reaction of acetic acid ($HC_2H_3O_2$) with water:

$$HC_2H_3O_2\ (aq) + H_2O\ (l) \leftrightarrow C_2H_3O_2^-\ (aq) + H_3O^+\ (aq).$$

Notice a subtle difference between the two equations. In the equation with HCl, the arrow only points to the right. This is because HCl is what we call a strong acid, meaning that when it dissolves in water, the result is a relatively high concentration of H_3O^+ ions. In fact, essentially 100 percent of the HCl molecules will react, leaving no neutral HCl molecules left in the solution.

Important Point

In a chemical equation when an arrow points in only one direction (→), the reaction only takes place in that direction. When an arrow points in both directions (↔), the reaction takes place in both directions.

In contrast, with acetic acid ($HC_2H_3O_2$), a weak acid, the arrow points both ways, meaning that the reaction proceeds in both directions. Actually, the favorable direction is to the left. Depending on the concentration of $HC_2H_3O_2$, only a few percent of the molecules actually react. A few $C_2H_3O_2^-$ and H_3O^+ ions do form, but the large majority of $HC_2H_3O_2$ remain in solution as neutral molecules. The result is a relatively very low concentration of H_3O^+ ions. The solution is still acidic, but only weakly so.

We see in both of these examples that the net result is the transfer of an H^+ ion from the acid to the base. (In these cases water plays the role of a base.) This gives us another—but completely equivalent—way of looking at the Brönsted-Lowry definition of an acid. Since the large majority of H^+ ions are just protons, to say that an acid transfers an H^+ ion is to say that an acid is an H^+ donor, or equivalently, a proton donor.

Chem Vocab

A strong acid completely dissociates into ions in aqueous solutions, resulting in a high concentration of H^+ ions.

A weak acid mostly still consists of undissociated molecules in aqueous solutions. Since only a few percent of the molecules have dissociated, the solution is only weakly acidic and the concentration of H^+ ions is low.

Bases

According to Brönsted and Lowry, a chemical species is a base if, when the species is added to water, the concentration of OH^- ions increases. We can use sodium hydroxide (NaOH) as an example. (Sodium hydroxide is the main ingredient in lye and in some drain cleaners.)

When NaOH is dissolved in water, the molecules dissociate into Na^+ and OH^- ions. Since nearly 100 percent of the molecules dissociate, the concentration of OH^- ions will be relatively high and NaOH is classified as a strong base.

The interaction between NaOH and water is difficult to express in an equation. Mostly, as the molecule dissociates into ions, the two ions each become surrounded by water molecules, separating the two ions from each other. We say that the ions have become hydrated.

A familiar example of a weak base is aqueous ammonia (NH_3), the main component of some window cleaners, which reacts with water as shown in the following equation:

$$NH_3\ (aq) + H_2O\ (l) \leftrightarrow NH_4^+\ (aq) + OH^-\ (aq).$$

Notice that the arrow again goes in both directions. Saying that ammonia is a weak base means that only a small percent of the ammonia molecules react with water. Some OH^- ions form. Although their concentration is low, the solution is still basic. Notice also (and very importantly) that it is not required that the species that is a base actually contain OH in its formula. The reaction does result in the increase in OH^- concentration, but in this case the OH^- ions come from the water molecules, not from the ammonia.

As before, we see in the example of ammonia and water that the net result is again the transfer of an H^+ ion from the acid to the base. (In this example water played the role of an acid.) Or, we can also look at it as the acceptance of H^+ by the base. This gives us another—but completely equivalent—way of looking at the Brönsted-Lowry definition of a base. Since the large majority of H^+ ions are just protons, to say that a base receives an H^+ ion from the acid is to say that a base is an H^+ acceptor, or equivalently, a proton acceptor.

In conclusion, we have two complementary definitions of Brönsted-Lowry acids and bases. From one point of view, we say that an acid is a substance that increases the concentration of H^+ (or H_3O^+) in a solution, and a base is a substance that increases the concentration of OH^-. From the other point of view, we say an acid is a proton

Chem Vocab

There are two equivalent ways to define Brönsted-Lowry acids and bases:

➤ An acid is a proton donor and a base is a proton acceptor.

➤ An acid is any substance that, when it is dissolved in water, results in an increase in the concentration of hydronium ions. A base is any substance that, when it is dissolved in water, results in an increase in the concentration of hydroxide ions.

donor, and a base is a proton acceptor. Either definition is a completely accurate way of thinking about acids and bases.

Strengths of Acids and Bases

The terms *strong* and *weak* acids and bases have already been used, but perhaps need to be explained more fully. *Strong* and *weak* refer to the degree or extent to which an acid or base ionizes in water. For all practical purposes, strong acids and strong bases dissociate completely. This means that only positive and negative ions are present in the solution; no undissociated acid or base molecules are present. The most common strong acids are hydrochloric acid (HCl, also called muriatic acid), nitric acid (HNO_3), and sulfuric acid (H_2SO_4). The most common strong bases are sodium hydroxide (NaOH) and potassium hydroxide (KOH). Because Cl^-, NO_3^-, Na^+, and K^+—and to a large extent, SO_4^{2-} —do not react with H^+ or OH^-, these ions are neither acidic nor basic and will behave as spectator ions in all acid-base reactions.

Weak acids and bases react to only a very slight extent with water. Although there will be a small number of H_3O^+ ions present in a solution of a weak acid, and a small number of OH^- ions present in a solution of a weak base, the concentration of ions is very low. Mostly only neutral acid or base molecules are present. Examples of common weak acids are carbonic acid (H_2CO_3), acetic acid ($HC_2H_3O_2$), phosphoric acid (H_3PO_4), and hydrogen sulfide (H_2S). As already mentioned, a common weak base is ammonia (NH_3), often called ammonium hydroxide (NH_4OH).

Chem Vocab

Neutralization means that relatively stronger acids and bases are combining to becoming relatively weaker acids and bases. In neutralization reactions, water—which is both a weak acid and a weak base—is often formed.

Neutralization Reactions

In a typical acid-base reaction, what mainly happens is the transfer of H^+ from the acid to the base. When equal amounts of a strong acid are mixed with a strong base, the result is neutralization. The resulting solution is much less acidic or basic than either of the original solutions; in fact, the resulting solution could be neutral, that is, neither acidic nor basic. The reaction between nitric acid and sodium hydroxide is an example of just such a reaction, as shown by the following equation:

$$HNO_3\ (aq) + NaOH\ (aq) \rightarrow Na^+\ (aq) + NO_3^-\ (aq) + H_2O\ (l).$$

Notice three features of this reaction:

> ➤ The arrow is pointing to the right. This means that if equal numbers of HNO_3 and NaOH molecules are mixed together, essentially 100 percent of both molecules will react.

> ➤ One of the products is water. Since water is the weakest acid present in an aqueous system, and the tendency of acids and bases is to react so as to form weaker acids and bases than the original ones, water is quite often a product of an acid-base reaction.

> ➤ Another product of the reaction is a salt, in this case, $NaNO_3$. "Na^+ (*aq*) + NO_3^- (*aq*)" could have been written as "$NaNO_3$ (*aq*)." Writing the two ions separately emphasizes the fact that they do not react or combine with each other in any way. This is the same thing as saying that $NaNO_3$ is soluble in water and is completely dissociated into ions. "$NaNO_3$ (*aq*)" is intended to say the same thing, but it requires you to remember that Na^+ and NO_3^- do not combine to form molecules.

Conjugate Acid-Base Pairs

According to the Brŏnsted-Lowry definition of acids and bases, when an acid donates a proton, what is left of the acid is itself a base—we call it the *conjugate base* of the acid. Likewise, when a base accepts a proton, it becomes an acid—the *conjugate acid* of the base. The idea of conjugate acid-base pairs is illustrated in the following reaction:

Chem Vocab

The conjugate base of an acid is what is left when H^+ is removed from the acid.

The conjugate acid of a base is what is formed when H^+ is added to the base.

$$H_3PO_4\ (aq) + OH^-\ (aq) \rightarrow H_2PO_4^-\ (aq) + H_2O\ (l).$$

$$\text{(acid)} + \text{(base)} \rightarrow \text{(base)} + \text{(acid)}$$

In this reaction, $H_2PO_4^-$ is the conjugate base of H_3PO_4, and H_2O is the conjugate acid of OH^-.

Table 9.1 lists several examples of acid-base conjugate pairs.

Table 9.1: Acid-Base Conjugate Pairs

Acids are listed from strongest to weakest, i.e., in order of decreasing strength. Bases are listed in order of increasing strength.

Name of Acid	Conjugate Acid	Conjugate Base	Name of Base
·Strongest Acids·			**·Weakest Bases·**
Perchloric	$HClO_4$	ClO_4^-	Perchlorate
Nitric	HNO_3	NO_3^-	Nitrate
Hydrochloric	HCl	Cl^-	Chloride
Sulfuric	H_2SO_4	HSO_4^-	Hydrogen sulfate
Hydronium	H_3O^+	H_2O	Water
·Weak Acids·			
Oxalic	$H_2C_2O_4$	$HC_2O_4^-$	Hydrogen oxalate
Hydrogen sulfate	HSO_4^-	SO_4^{2-}	Sulfate
Phosphoric	H_3PO_4	$H_2PO_4^-$	Dihydrogen phosphate
Hydrofluoric	HF	F^-	Fluoride
Hydrogen oxalate	$HC_2O_4^-$	$C_2O_4^{2-}$	Oxalate
Acetic	$HC_2H_3O_2$	$C_2H_3O_2^-$	Acetate
Carbonic	H_2CO_3	HCO_3^-	Bicarbonate
Hydrogen sulfide	H_2S	HS^-	Hydrogen sulfide
Dihydrogen phosphate	$H_2PO_4^-$	HPO_4^{2-}	Hydrogen phosphate
Ammonium	NH_4^+	NH_3	Ammonia
Bicarbonate	HCO_3^-	CO_3^{2-}	Carbonate
Hydrogen phosphate	HPO_4^{2-}	PO_4^{3-}	Phosphate
Hydrogen sulfide	HS^-	S^{2-}	Sulfide
Water	H_2O	OH^-	Hydroxide
·Weakest Acids·			**·Strongest Bases·**

Notice that some of the acids in Table 9.1 also appear in the column of bases, and vice versa. Examples include the bicarbonate ion (HCO_3^-) and water (H_2O). Since a substance like HCO_3^- is both an acid *and* a base, HCO_3^- is both a proton donor and a proton acceptor. Substances that are both proton donors and proton acceptors are said to be *amphiprotic* (the prefix "amphi–" means "both").

Also notice that some acids contain more than one hydrogen ion in their formula. A general term for this kind of acid is *polyprotic*. Specifically, a substance like sulfuric acid (H_2SO_4) is said to be *diprotic* and a substance like phosphoric acid (H_3PO_4) is said to be *triprotic*.

The pH Scale

The acidity of a solution is expressed in terms of the concentration of hydrogen (or hydronium) ions in the solution. Typically, concentrations of H^+ vary from 1 M (or more) for very acidic solutions to 10^{-14} M (or less) for very basic solutions. (See Chapter 5 to review units of concentration.)

Expressing exponential numbers can be cumbersome and difficult to compare to each other. Chemists, therefore, instead use an acidity scale that was invented by a chemist in a bottling factory whose job it was to keep track of the acidity of the beverages being bottled in the factory. The symbol *pH* is derived from H for H^+ ions and p for the power of 10 in the molarity of the H^+ ion. Because the exponents are usually negative numbers, the sign is changed. Thus, the pH of a solution is defined as $pH = -\log_{10} [H^+]$.

Chem Vocab

Amphiprotic means that something is both a proton donor (an acid) *and* a proton acceptor (a base).

Chem Vocab

A *polyprotic* acid is an acid that contains more than one ionizable hydrogen ion. Examples include H_2CO_3, H_2SO_4, both of which are *diprotic*, and H_3PO_4, which is *triprotic*.

Chem Vocab

The pH of a solution is a measure of the solution's acidity. On a scale commonly ranging from 1 to 14, solutions with low pHs are acidic, solutions with high pHs are basic, and solutions with a pH of 7 are neutral.

Important Point

In the mathematics of logarithmic and exponential numbers, *log* means "logarithm." The subscript *10* means "base 10." Anytime we need to evaluate an expression of the form $\log_{10}(10^a)$, were *a* is a number, the answer is $\log_{10}(10^a) = a$.

For example, $\log_{10}(10^4) = 4$, and $\log_{10}(10^{-5}) = -5$.

Table 9.2 shows the pHs of solutions for concentrations of H^+ that are powers of 10.

Table 9.2: Concentration of H^+ and pH

$[H^+]$ (M)	1.0	10^{-1}	10^{-2}	10^{-3}	10^{-4}	10^{-5}	10^{-6}	10^{-7}
pH	0	1	2	3	4	5	6	7

Very Acidic   ← Acidity Increases   Neutral

$[H^+]$ (M)	10^{-8}	10^{-9}	10^{-10}	10^{-11}	10^{-12}	10^{-13}	10^{-14}	10^{-15}
pH	8	9	10	11	12	13	14	15

Alkalinity Increases →   Very Alkaline

The numbers in Table 9.2 may be interpreted as follows:

➤ A pH of 7 is a neutral solution. In that situation, $[H^+] = [OH^-]$.

➤ A pH less than 7 is an acidic solution. Acidic solutions have high concentrations of H^+.

➤ A pH greater than 7 is a basic solution. Basic solutions have high concentrations of OH^-.

➤ Solutions of strong acids (e.g. HCl) generally have pHs of 2 or less. A pH less than 0 would be a negative number, and corresponds to a very high concentration of H^+ ions.

➤ Solutions of weak acids (e.g. acetic acid) generally have pHs in the range of 3–6.

➤ Solutions of weak bases (e.g. ammonia) generally have pHs in the range of 8–11.

➤ Solutions of strong bases (e.g. NaOH) generally have pHs of 12–14.

Table 9.3 shows the pHs of common products, which should give you an idea of what kinds of household products are acids or bases.

Table 9.3: pHs of Common Solutions

pH	Examples of Solutions
0	Battery acid
1	Stomach acid
2	Lemon juice, vinegar
3	Orange juice, carbonated beverages
4	Tomato juice
5	Unpolluted rainwater
6	Urine, saliva
7	Pure water
8	Seawater
9	Baking soda
10	Milk of magnesia
11	Ammonia cleaning solutions
12	Soap solutions
13	Oven cleaner
14	Lye

Conclusion

Many of the substances we encounter in our daily lives are acids, bases, or salts. Many fruits and vegetables are acidic because of the citric acid in them. Vinegar and lemon juice are acidic. Many household substances, including baking soda and window cleaners, are bases. Ionic compounds that are not acids or bases are classified as salts, of which table salt (NaCl) is certainly the most familiar.

There are also many substances, however, that are not acids, bases, or salts. Substances like hydrogen peroxide (H_2O_2) and household bleach (5 percent sodium hypochloride, NaClO) are classified as *oxidizing* agents. Oxidation-reduction reactions are the subject of the next chapter.

CHAPTER 10

Oxidation-Reduction Reactions

In This Chapter

➤ Oxidation numbers

➤ Combustion reactions

➤ Activity series of the metals

➤ Photosynthesis and Respiration

An important property of an element is the oxidation state in which it exists in a compound or ion. We begin with learning how to assign oxidation numbers, part of a bookkeeping system that helps us keep track of what goes on in an important class of chemical reactions—oxidation-reduction reactions.

Chem Vocab

Oxidation number and oxidation state mean the same thing. In the case of simple ions, an element's oxidation state is the same as the charge on the ion.

Rules for Assigning Oxidation Numbers

The rules for assigning oxidation numbers are found in Table 10.1.

Table 10.1: Rules for Assigning Oxidation Numbers

Chemical Species	Rule	Examples
1. Neutral elements	Oxidation number = 0	Na, Cu, P; each atom in H_2, O_2, O_3, P_4, S_8
2. Simple ions: ions that contain only 1 atom	Oxidation number = charge on the ion	Cu = +2 in Cu^{2+} S = –2 in S
The next rules apply to compounds or ions:		
3. Alkali metals	Oxidation number always = +1	Li in LiCl, Li_2O, LiH; K in KOH, K_2CO_3, KNO_3
4. Alkaline earths	Oxidation number always = +2	Mg in $Mg(NO_3)_2$, Mg_3N_2, Ca in $CaSO_4$, $Ca_3(PO_4)_2$
5. Hydrogen atoms	Oxidation number = +1 if H is bonded to a nonmetal; = –1 if H is bonded to a metal	H = +1: H_2O, HCl, CH_4, $HC_2H_3O_2$, NaOH H = –1: LiH, CaH_2
6. Oxygen atoms	Oxidation number = –1 in a "peroxide"; = –2 in all other molecules or ions.	O = –1 in H_2O_2, Na_2O_2 O = –2 in H_2O, OH^-, CH_4O, $CaCO_3$, Fe_2O_3, $KClO_3$
7. Elements in compounds not covered by rules 3–6	The sum of all the oxidation numbers must = 0.	C = +4 in CO_2; C = –4 in CH_4; N = +5 in HNO_3; S = +6 in SO_3; Fe = +3 in Fe_2O_3
8. Elements in complex ions not covered by rules 3–6	The sum of all the oxidation numbers must = charge on the ion.	S = +6 in SO_4^{2-}; Cl = +1 in ClO^-; Mn = +7 in MnO_4^-; Hg = +1 in Hg_2^{2+}

Let us look at several examples to make sure you understand how to assign oxidation numbers.

➤ Example 1 – What are the oxidation numbers of all the elements in $NaClO_4$?

Solution – By rule 3, Na = +1. By rule 6, each O = –2. To find the oxidation number of Cl (call it *X*), use rule 7:

$+1 + X - 4(-2) = 0$. Solve for X. Result: $X = +7$. So, the oxidation number (or oxidation state) of Cl is +7.

➤ Example 2 – What are the oxidation states of carbon and oxygen in $C_2O_4^{2-}$?

Solution – By rule 6, each $O = -2$. Use rule 8 to find the oxidation state of each C atom (*X*): $2X + 4(-2) = -2$ (the charge on the ion). Solve for X. Result: $X = +3$, which is the oxidation state of each C atom.

➤ Example 3 – What are the oxidation states of iron and oxygen in FeO and Fe_2O_3?

Solution – In each one of these, rule 6 tells us that each O atom is –2. Rule 7 tells us that the sum of the oxidation states is 0. Therefore, for FeO, $Fe = +2$. For Fe_2O_3, each $Fe = +3$.

Sometimes the rules cannot be applied unambiguously to a chemical species. Consider KSCN. By rule 3, the oxidation state of K is +1. However, none of the other three elements—S, C, or N—is covered by the rules. We will not try to assign oxidation states in cases like this. Even professional chemists might disagree as to what oxidation states these elements are in.

There are some general statements that can be made about oxidation states:

➤ An element in an even-numbered column tends to have even-numbered oxidation states.

➤ An element in an odd-numbered column tends to have odd-numbered oxidation states.

Be careful, though. There are exceptions to the even-odd rules. For example, Mn is in column VII (an odd number) but one of its common compounds is MnO_2, in which the oxidation state of Mn is +4 (an even number). Another example occurs with Cu, which is in column I (an odd number), but copper's most common ion is Cu^{2+}, in which Cu is in the +2 state (an even number).

Important Point

Iron in the +2 oxidation state is the ferrous ion. Iron in the +3 oxidation state is the ferric ion.

Important Point

In compounds or ions, each atom is assigned an oxidation state. In H_2O, for example, the oxygen atom is in the –2 state. Each hydrogen atom is in the +1 state. It is incorrect to say that hydrogen is in a +2 state.

➤ When families, or groups, of elements are numbered using Roman numerals (I, II, III, etc.), the family number almost always is the highest possible oxidation state possible for an element in that column.

Again, there are a few exceptions, although not many. Two notable exceptions are that both Cu and Au are in column I, but Cu readily forms a +2 ion and Au forms a +3 ion. All of these rules are convenient rules of thumb, however, if not observed too stringently.

➤ In compounds or ions, metals can *only* have positive oxidation states. Any exceptions would be too exotic to include in this book.

➤ Nonmetals can exist in either positive *or* negative oxidation states. An example is carbon. In CH_4, C is in the –4 state. In CO_2, C is in the +4 state. (In other compounds, carbon can also be in oxidation states lying between those two extremes.) Another example is chlorine. In NaCl, Cl is in the –1 state. In $HClO_4$, Cl is in the +7 state. (Like carbon, chlorine can also assume oxidation states lying between those extremes.)

➤ The range of oxidation states elements can assume is one of the chemical properties of elements. Since the chemical properties of metalloids tend to resemble the chemical properties of nonmetals more than they do the chemical properties of metals, metalloids (e.g. B, Si, and As) can often assume both positive and negative oxidation states.

➤ The noble gases present a small problem when using Roman numerals. Since noble gases tend to exist only as single gaseous atoms, their oxidation numbers should be zero. However, there is no zero in Roman numerals. Some people get around this problem by labeling the noble gases as column VIII. Others just label it as column 0, recognizing that 0 is not a Roman numeral.

Important Point

An important distinction between metals and nonmetals (and metalloids) is the kinds of oxidation states in which they can be found. In compounds and ions, metals can *only* be in positive oxidation states. Non-metals and metalloids can be in either positive or negative oxidation states.

Why Oxidation Numbers Are Important

Why are oxidation numbers important? The answer is that they help us to understand what is taking place in an important class of chemical reactions—oxidation-reduction reactions (or redox reactions for short).

Processes such as the combustion of fuels, the refinement of ores into their component metals, and the corrosion of metals have been among the most familiar chemical reactions since ancient times. The nature of these reactions, however, remained a complete mystery throughout the Middle Ages and well into modern times. It was not until the discovery of oxygen in the 1770s that processes as familiar as burning wood could be explained.

Rather quickly, chemists of the late 18th century were finally able to associate combustion processes with the reaction of fuels—wood, coal, peat, oil—with oxygen found in the atmosphere. Thus, gaining oxygen atoms came to be called oxidation and the loss of oxygen atoms reduction.

Later, it was realized that combustion reactions and the reactions of metals are just examples of more general processes that do not necessarily require the presence of oxygen. The terms *oxidation* and *reduction*, therefore, have been generalized to include a broader class of chemical reactions.

Historical Note

After oxygen was discovered in the late 1700s and chemists recognized the role it plays in combustion reactions, the term *oxidation* was coined to mean "combining with oxygen." During the 1800s chemists began to realize that there are a lot of oxidation processes that do not involve oxygen, but the term *oxidation* persisted.

In a redox reaction the oxidation numbers of two elements will change—the oxidation number of one element will increase, and the oxidation number of another element will decrease. An increase in oxidation number is called oxidation, and a decrease is called reduction. Oxidation and reduction must occur in pairs; one cannot occur without the other one occurring also.

Elements in nature can be found in different oxidation states. For example, the mineral cuprite has the formula Cu_2O. By rule 6, copper is in the +1 oxidation state. Malachite, on the other hand, has the formula $Cu_2CO_3(OH)_2$. Given that carbon is in the +4 state, each oxygen is in the −2 state, and each hydrogen is in the +1

Chem Vocab

Oxidation always means an increase in oxidation number.

Reduction always means a decrease in oxidation number.

state, copper must be in the +2 oxidation state. Similarly, hematite has the formula Fe_2O_3. Iron is in the +3 oxidation state. Magnetite has the formula Fe_3O_4. At first glance it looks like iron is in the 8/3 oxidation state. The formula Fe_3O_4, however, can also be written $FeO\cdot Fe_2O_3$. We see that the first iron atom is in the +2 state, but the other two iron atoms are in the +3 state. Magnetite illustrates the fact that the same compound can sometimes contain atoms of the same element that are in different oxidation states.

Combustion Reactions

Combustion reactions are among our most important chemical reactions. Our technological, energy-intensive society relies heavily on the combustion of fossil fuels as a source of energy. In the combustion of hydrocarbons, such as the substances found in natural gas or petroleum, the fuel combines with oxygen gas (O_2) from the atmosphere to make carbon dioxide (CO_2) and water (H_2O).

Important Point

Because our society depends so heavily on energy derived from the burning of fossil fuels, combustion reactions are among our most important examples of oxidation-reduction processes.

For example, the principal component of natural gas is methane, CH_4. The chemical equation for the combustion of CH_4 is the following:

$$CH_4 \text{ (g)} + 2\,O_2 \text{ (g)} \rightarrow CO_2 \text{ (g)} + 2\,H_2O \text{ (g)}.$$

In CH_4, C is in the –4 state; in CO_2, it is in the +4 state. A change from –4 to +4 is an increase. Therefore, we would say that carbon has been oxidized. In O_2, oxygen is in the 0 oxidation state because it is a neutral element; in CO_2 and H_2O, it is in the –2 state. Going from 0 to –2 is a decrease and we would say that oxygen has been reduced. This will be true of *all* combustion reactions; an element in the fuel will be oxidized and oxygen will be reduced. Thus, combustion is an oxidation-reduction reaction.

The Activity Series of the Metals

In nature most metals have a greater tendency to exist in ores than as neutral metals. Therefore, when metals are mined they are already in oxidized states. To be useful in industry, they must be reduced to the neutral metals. Iron, for example, is most commonly found in the minerals limonite (FeO), hematite (Fe_2O_3), magnetite (Fe_3O_4), or in iron ores mixed with other metals. With limonite, metallic iron is produced in a blast furnace by reacting FeO with carbon monoxide (CO) at elevated temperatures, as shown:

$$FeO\ (l) + CO\ (g) \rightarrow Fe\ (l) + CO_2\ (g).$$

In this case, we see that iron has been reduced from the +2 state to the 0 state and carbon has been oxidized from the +2 to the +4 state.

Metals may also be oxidized by reacting them with substances other than oxygen gas. For example, when sterling silverware tarnishes, silver reacts with traces of hydrogen sulfide (H_2S) in the atmosphere, as shown:

$$2\ Ag\ (s) + H_2S\ (g) \rightarrow Ag_2S\ (s) + H_2\ (g).$$

Silver sulfide (tarnish) is black. In this case, silver is oxidized and hydrogen is reduced. Polishing tarnished silverware removes the sulfur and reduces silver to the neutral metal again.

Neutral metals of one element can also react with metal ions of other elements. For example, zinc metal reacts with copper ions, as shown:

$$Zn\ (s) + Cu^{2+}\ (aq) \rightarrow Zn^{2+}\ (aq) + Cu\ (s).$$

Zinc has been oxidized and copper ions have been reduced.

Sometimes, a metal can also react with one of its own ions, as shown for Sn and Sn^{4+}:

$$Sn\ (s) + Sn^{4+}\ (aq) \rightarrow 2\ Sn^{2+}\ (aq).$$

In this case the tin metal and the tin (IV) ion begin in different oxidization states; tin metal has been oxidized and the tin (IV) ion has been reduced. The reverse reaction is an example of *disproportionation*, a term which describes a process in which the atoms of an element all begin in the same oxidation state, but some are oxidized while some are reduced. Thus, in the reaction

$$2\ Sn^{2+}\ (aq) \rightarrow Sn\ (s) + Sn^{4+}\ (aq),$$

one Sn^{2+} ion is oxidized and one ion is reduced.

Chem Vocab

Disproportionation means that in a chemical reaction, some of the atoms of an element are oxidized and other atoms of the same element are reduced.

Important Point

Because metals tend to be very reactive, most metals are mined from ore deposits in which the metals occur in compounds. Only a few metals are unreactive enough that they can be found in the pure state as native metals. Examples of native metals include copper, silver, gold, and platinum.

Metals and metalloids often are arranged in an *activity series*, which shows the metals which oxidize most easily at the top of the list, and the metals which most resist oxidation at the bottom. An example of an activity series of metals is shown in Table 10.2.

Chem Vocab

An active metal is one that is easily oxidized.

The activity series lists the metals in order of ease with which they can be oxidized.

Table 10.2: Activity Series of the Metals

Most Active	RuLi	
	Rb	
	K	
	Cs	These elements react with cold water, liberating H_2.
	Ba	
	Sr	
	Ca	
	Na	
	Mg	
	Be	
	Ti	These elements react with steam, liberating H_2.
	Al	
	Zr	
	Mn	
	V	
	Cr	
	Zn	
	Ga	
	W	
	Ga	
	W	
	Mo	These elements react with HCl and H_2SO_4, liberating H_2.
	Fe	
	Sn	
	Cd	
	Co	
	Ni	
	Pb	
	As	
	Sb	
	Bi	
	Cu	These elements react with HNO_3, liberating NO or NO_2.
	Te	
	Se	
	Os	
	Ag	
	Hg	
	Pd	
	Ru	
	Rh	These elements react with *aqua regia*, liberating NO or NO_2
	Pt	
	Ir	
Least Active	Au	

*a mixture of concentrated HCl and HNO_3

We may draw some general conclusions from the activity series:

➤ The alkali metals are easiest to oxidize, followed by the alkaline earths. These metals, in fact, readily react (sometimes explosively) with cold water to form their ions. We can draw the conclusion, therefore, that we would never find alkali or alkaline earth elements in their neutral form in nature.

➤ Other metals that oxidize easily include Zn and Al. These metals easily dissolve in dilute solutions of weak acids, and sometimes in dilute solutions of bases. (If a metal dissolves in either acid or base, it is said to be amphoteric. Aluminum is an example.)

➤ Metals like Fe, Ni, Sn, and Pb dissolve less easily than Zn or Al do, but will dissolve in strong acids like hydrochloric (HCl) or sulfuric (H_2SO_4) acids.

➤ Metals that occupy positions below Pb in the activity series do not dissolve in HCl or H_2SO_4. Metals like Cu, Hg, and Ag require nitric acid (HNO_3) which is a more powerful oxidizing agent than the first two strong acids.

➤ Metals at the bottom of series are *very* difficult to dissolve or oxidize by other reactions, especially Au and the platinum group metals (Pt, Pd, Rh, Ir, Ru, and Os). They generally require a very powerful oxidizing agent, *aqua regia* (literally, "royal water" in reference to dissolving gold), which is a mixture of concentrated HCl and HNO_3.

➤ Metals at the bottom of the series may be found as native metals, that is, as the neutral metals. Gold nuggets are familiar examples. Other elements sometimes found as native metals include Cu, Ag, Pt, and Bi.

➤ Since ancient times, metals at the bottom of the activity series have been used for coinage, jewelry, and industrial applications. In fact, the family that includes Cu, Ag, and Au is often called the coinage metals.

Important Point

The metals located in the first two columns of the periodic table are called *active* metals because they are so extremely reactive with water, oxygen, and other common chemicals.

Corrosion

Iron, in the form of its principal alloy, steel, is one of civilization's major structural materials. In contact with water, oxygen, and other chemicals in the environment, however, iron *corrodes* (oxidizes) easily. The familiar reddish-brown substance rust (Fe_2O_3) forms under wet conditions according to the following reactions:

$$4\ Fe\ (s) + 3\ O_2\ (g) + 6\ H_2O\ (l) \rightarrow 4\ Fe(OH)_3\ (s),$$

followed by:

$$4\ Fe(OH)_3\ (s) \rightarrow 2\ Fe_2O_3\ (s) + 6\ H_2O\ (l).$$

In the neutral metal, Fe is in the 0 oxidation state; in Fe_2O_3, it is in the +3 state. Therefore, iron has been oxidized. In oxygen gas, oxygen atoms are in the 0 oxidation state; in $Fe(OH)_3$ or Fe_2O_3, oxygen atoms are in the –2 state. Therefore, oxygen has been reduced.

Important Point

Most metals are found in nature in ores in which the metals are in oxidized states. Ores are not useful as structural materials, so they have to be reduced to their neutral metals. This process is very energy intensive and is a major source of energy consumption in industry.

Photosynthesis and Respiration

Two of the most biologically important chemical reactions that take place are photosynthesis and respiration. *Photosynthesis* means "light" plus "making chemicals." Using energy in the form of sunlight, green plants manufacture sugars. Plants are producers, the source of nearly all food on Earth.

Photosynthesis takes place in the presence of chlorophyll in the cells of plants. The chemical equation for photosynthesis is the following:

$$6\ CO_2\ (g) + 6\ H_2O\ (l) \rightarrow C_6H_{12}O_6\ (s) + 6\ O_2\ (g).$$

$C_6H_{12}O_6$ is glucose, an important sugar. Glucose can be turned into starch for food. Thus, photosynthesis results in the production of food for a plant's own growth as well as food for animals. Equally important, photosynthesis is the source of oxygen gas, which is essential to animals. In CO_2, carbon starts out in the +4 oxidation state. In $C_6H_{12}O_6$, carbon has been reduced to the 0 oxidation state. In H_2O, oxygen starts out in the –2 state. In O_2, oxygen has been oxidized to the 0 state. Thus, photosynthesis is an oxidation-reduction reaction. It is also an endothermic process since energy must be absorbed for the reaction to take place.

Important Point

Photosynthesis and respiration are just the opposite of each other. During photosynthesis energy is absorbed and CO_2 and H_2O are converted into sugar and O_2. During respiration energy is released and sugar combines with O_2 to make CO_2 and H_2O.

Chemically speaking, respiration is just the opposite of photosynthesis. Plants and animals burn glucose in the mitochondria of their cells, making CO_2 and H_2O, as shown:

$$C_6H_{12}O_6 \,(s) + 6\,O_2\,(g) \rightarrow 6\,CO_2\,(g) + 6\,H_2O\,(l)$$

Respiration is a combustion process and is therefore exothermic, although combustion in the cells of living organisms takes place very, very slowly by a complicated series of steps. Otherwise, so much heat would be produced that it would lead to the death of the organism. As in the combustion of any carbon-based fuel, however, carbon is being oxidized and oxygen gas is being reduced.

Conclusion

The concepts introduced in this chapter are used throughout chemistry. The applications of oxidation-reduction reactions appear throughout science and industry. Two of these applications define the field of electrochemistry, which is the subject of the next chapter.

CHAPTER 11

Electrochemical Processes

In This Chapter

➤ History of Electricity

➤ Batteries

➤ Electrolysis

Most of the time, when we talk about chemical reactions, we are talking about situations in which two or more chemicals are mixed together in direct contact with each other and a reaction takes place. There is a special kind of chemical reaction, though, in which the reacting chemical species do not come into direct contact with each. Instead, an electrochemical reaction takes place in which an electrical circuit connects the reacting species and current flows between them. To begin the chapter, however, we'll look at the history of electrochemical reactions.

Chem Vocab

An electrochemical reaction is a chemical reaction in which an electrical current flows through the system.

Historical Background

Before we get into the chemistry, let us learn something about the names of the people who are commemorated in the terms we use. The galvanic cell is named after Luigi Galvani (1737–98), an Italian biologist and medical doctor who studied the effects of electrical currents on living organisms such as frogs.

The unit of electrical potential, the volt, is named after Alessandro Volta (1745–1827), an Italian physicist whom history credits with the inventions of the battery and the capacitor (a device that stores electrical charge).

Chem Vocab

The ampere is the unit of electrical current.

The volt is the unit of electrical potential.

The ohm is the unit of electrical resistance.

The unit of electrical current, the ampere, is named after André Marie Ampère (1775–1836), a French mathematician and physicist whom history credits with the invention of the galvanometer (a device that measures electrical currents).

The unit of electrical resistance, the ohm, is named after Georg Simon Ohm (1787–1854), a German physicist who discovered the law that bears his name. Ohm's Law states that the potential difference (in volts) measured in an electrical circuit equals the product of the current (in amperes) times the resistance (in ohms).

Galvanic Cells

A galvanic cell has two electrodes: the anode, which is negatively charged and is the site at which the oxidation half-reaction takes place; and the cathode, which is positively charged and is the site at which reduction occurs. The oxidation-reduction reaction taking place in a galvanic cell occurs spontaneously, which means that energy does not need to be supplied to make the reaction occur. Instead, a galvanic cell produces energy in the form of an electrical current and can be used to do electrical work, such as running an electric motor or lighting a lightbulb. When a galvanic cell is packaged and sold commercially, we call it a battery.

Important Point

In both galvanic and electrolytic cells, oxidation always occurs at the anode and reduction always occurs at the cathode. However, the polarities of the electrodes are different in the two kinds of cells. In a galvanic cell the cathode is + and the anode is –. In an electrolytic cell the cathode is – and the anode is +.

The Standard Hydrogen Potential

In any galvanic cell an electrical potential difference exists between the two electrodes. Potential difference is measured in volts. The two sides of a galvanic cell are called half-cells. Although neither the oxidation nor the reduction half-reactions can take place by themselves, we can think of the two half-cells as though they were two batteries connected in series.

A series circuit is the one used in flashlights and other devices that use batteries—if two or more batteries are used together, they are connected + to –. (If circuit elements are connected + to + and – to –, it is called a parallel circuit. We can ignore parallel circuits in this discussion because galvanic cells always have series circuits.)

When we connect a voltmeter to the electrodes of a galvanic cell, we are measuring the electrical potential difference of the total cell, which is the sum of the potential differences of the two half-cells. Since we cannot measure the potential difference of a half-cell by itself, chemists have chosen to assign the standard hydrogen electrode (SHE) a potential difference of 0.00 v. The standard hydrogen electrode consists of a platinum electrode immersed in a 1 M HCl solution with hydrogen gas bubbling into the solution at a pressure of 1 atmosphere (atm).

If the hydrogen electrode is the anode, then the half-reaction taking place is

$$H_2\,(g) \rightarrow 2\,H^+\,(aq) + 2\,e^-.$$

If the hydrogen electrode is the cathode, then the reaction taking place is just the opposite:

$$2\,H^+\,(aq) + 2\,e^- \rightarrow H_2\,(g).$$

Either way, we define the standard potential, symbolized E°, as 0.00 v.

Important Point

The symbol E is short for electromotive force, which is another way of thinking about the potential difference between the electrodes. The superscript ° means "standard conditions," which are gas pressures of 1 atm and ion or molecule concentrations of 1 M. The value of E° changes if pressures or concentrations change.

Standard Potentials

We can measure the standard potentials for all other oxidation or reduction half-reactions by constructing galvanic cells in which SHE is one half-cell and the reaction whose E° we want to measure is the other half-cell. For example, when we are measuring E° for the half-reaction

$$Zn\ (s) \rightarrow Zn^{2+}\ (aq) + 2\ e^-,$$

the total reaction is

$$Zn\ (s) + 2\ H^+\ (aq) \rightarrow Zn^{2+}\ (aq) + H_2\ (g).$$

Since +0.76 v is the sum of the E°'s for the two half-reactions, and E° for SHE is 0.00 v, we conclude that E° for the oxidation of zinc metal is +0.76 v.

If we construct a galvanic cell to measure E° for the oxidation of Cu, we have a cell in which the total reaction that occurs is

$$Cu^{2+}\ (aq) + H_2\ (g) \rightarrow Cu\ (s) + 2\ H^+\ (aq).$$

Notice an extremely important difference between this cell and the one using zinc. In the cell containing zinc,, zinc was the anode, which meant that zinc metal was being oxidized to zinc ions. Copper, however, is the cathode, which means that copper *ions* are being reduced to copper metal. That means that E° = +0.34 v is the standard potential for the half-reaction

$$Cu^{2+}\ (aq) + 2\ e^- \rightarrow Cu\ (s).$$

If we want to know E° for the oxidation of copper metal, we must reverse this half-reaction, which means that we change the sign on E°. E° is –0.34 v for the half-reaction

$$Cu\ (s) \rightarrow Cu^{2+}\ (aq) + 2\ e^-.$$

Important Point

The reverse of any oxidation half-reaction is a reduction half-reaction, and vice versa. When we reverse the direction of the half-reaction, we change the sign on E°.

The Meaning of Standard Potentials

You are probably wondering what difference does it make if E° is + or –? What do + and – values mean? The answer is that the standard potential for any battery is always a positive number. Positive values of E° mean that the reactions take place spontaneously. Galvanic cells are a source of energy—they can be made to do useful electrical work.

Negative values of E^o mean that the reactions do not take place by themselves. An external source of energy needs to do electrical work to make the reaction take place. In that case, we say the reaction is taking place in an electrolytic cell instead of a galvanic cell. The difference between a cell that does electrical work and a cell that needs electrical work to be done on it is a crucial distinction.

We usually do not make galvanic cells using the standard hydrogen electrode. Using the two examples from above, suppose we make a galvanic cell that has the zinc half-reaction in one half-cell and the copper half-reaction in the other one. Then the standard potentials are:

$$Zn\ (s) \rightarrow Zn^{2+}\ (aq) + 2\ e^- \quad E^o = +0.76\ v$$

$$Cu^{2+}\ (aq) + 2\ e^- \rightarrow Cu\ (s) \quad E^o = +0.34\ v.$$

Together, the combined reaction is

$$Zn\ (s) + Cu^{2+}\ (aq) \rightarrow Zn^{2+}\ (aq) + Cu\ (s).$$

The total standard potential is found by adding the two values of E^o together:

$$E^o = +0.76\ v + 0.34\ v = +1.10\ v.$$

Values of E^o have been measured for a large number of half-reactions. Table 11.1 lists the values of E^o for only a few common half-reactions. Notice that all the half-reactions are reduction half-reactions. They could just as well be oxidation half-reactions, but the common practice is to list reduction half-reactions.

Table 11.1 – Table of Standard Reduction Potentials (25˚C)

$Ag^+ + e \rightarrow Ag^\circ$	+0.80 V
$Al^{3+} + 3e \rightarrow Al^\circ$	−1.66
$Au^{3+} + 3e \rightarrow Au^\circ$	+1.40
$Bi^{3+} + 3e \rightarrow Bi^\circ$	+0.20
$Br_2 + 2e \rightarrow 2\ Br^-$	+1.09
$Ca^{2+} + 2e \rightarrow Ca^\circ$	−2.87
$Cl_2 + 2e \rightarrow 2\ Cl^-$	+1.36
$Co^{2+} + 2e \rightarrow Co^\circ$	−0.28
$Cr^{3+} + 3e \rightarrow Cr^\circ$	−0.74
$Cs^+ + e \rightarrow Cs^\circ$	−2.92
$Cu^{2+} + 2e \rightarrow Cu^\circ$	+0.34
$F_2 + 2e \rightarrow 2\ F^-$	+2.87
$Fe^{2+} + 2e \rightarrow Fe^\circ$	−0.44
$Fe^{3+} + e \rightarrow Fe^{2+}$	+0.77
$2\ H^+ + 2e \rightarrow H_2$	0.00
$2\ H_2O + 2e \rightarrow H_2(g) + 2\ OH^-$	−0.83
$I_2 + 2e \rightarrow 2\ I^-$	+0.54
$Li^+ + e \rightarrow Li^\circ$	−3.05
$Mg^{2+} + 2e \rightarrow Mg^\circ$	−2.37
$Na^+ + e \rightarrow Na^\circ$	−2.71
$Ni^{2+} + 2e \rightarrow Ni^\circ$	−0.25
$NO_3^- + 4\ H^+ + 3e \rightarrow NO(g) + 2\ H_2O$	+0.96
$O_2(g) + 2\ H_2O + 4e \rightarrow 4\ OH^-$	+0.40
$PbO_2(s) + 4\ H^+ + SO_4^{2-} + 2e \rightarrow PbSO_4(s) + 2\ H_2O$	+1.69
$PbSO_4(s) + 2e \rightarrow Pb^\circ + SO_4^{2-}$	−0.36
$Zn^{2+} + 2e \rightarrow Zn^\circ$	−0.76

Not all oxidation-reduction reactions use neutral metals. In those cases some inert, conducting electrode is used. In SHE, platinum is used. A less expensive substitute is a rod of graphite.

Commercial Batteries

Several commercial batteries should be familiar. Table 11.2 lists some examples.

Table 11.2 – Commercial Batteries

Kind of Battery	Reaction	+ and – electrodes	E°
Flashlight (AA, AAA, C, D) Non-alkaline	$Zn + 2\,NH_4Cl + 2\,MnO_2 \rightarrow Zn(NH_3)_2Cl_2 + H_2O + Mn_2O_3$; uses a $ZnCl_2/NH_4Cl$ paste in water as the electrolyte	MnO_2 in graphite $= +$ $Zn = -$	1.5 v
Alkaline (AA, AAA, C, D)	Same as non-alkaline batteries, but with a potassium hydroxide electrolyte	same	1.5 v
Ni-Cad Rechargeable	$NiO(OH) + 2\,Cd + H_2O \rightarrow Ni(OH)_2 + Cd(OH)_2$	$NiO(OH) = +$ $Cd = -$	1.2 v
Lead Storage (Motor Vehicles)	$Pb + PbO_2 + 2\,H_2SO_4 \rightarrow 2\,PbSO_4 + 2\,H_2O$	$PbO_2 = +$ $Pb = -$	2.0 v per cell (car batteries have 6 cells = 12 v)
Hydrogen Fuel Cell	$2\,H_2 + O_2 \rightarrow 2\,H_2O$	$H_2 = +$ $O_2 = -$ (metal electrodes)	0.40 v

Ohm's Law

The value of E° for a galvanic cell does not tell us very much about the current produced by that cell. For example, eight AA batteries connected in series generate 12 v, as does a car battery with its six cells. However, the current produced by a car battery is huge compared to the current produced by any number of AA batteries. The difference is due to Ohm's law.

We can write Ohm's law as

$$V = IR, \text{ or } I = \frac{V}{R},$$

where V = potential difference (in volts), I = current (in amperes), and R = resistance (in ohms). From inspection, we see that batteries, or a set of batteries connected in series, can have the same voltage, but produce very different currents if their resistances differ. One factor that affects resistance is the cross-sectional area through which current has to pass. Small cross-sectional areas have high resistance, and large areas have low resistance.

AA, AAA, C, and D batteries are made as thin cylinders with very small cross sectional areas through which the current passes, therefore, their resistances are high and their currents are low. Car batteries are made of large Pb and PbO_2 plates with very large surface areas, therefore, their resistances are low and their currents are very high. Car batteries must generate high currents; their function is to provide the energy needed to start the engine. A lot of current is required to do that.

Important Point

Ohm's law states that current = voltage divided by resistance:

$I = \frac{V}{R}$. Battery voltages generally range from about 1.5 to 12 volts. Resistances, however, vary greatly. High resistance means low current; the battery's electrochemical reaction takes place very slowly. Low resistance means high current; the battery's reaction takes place rapidly.

The requirement for high currents means several other differences between car batteries and flashlight batteries. Lead is a very dense metal. The use of large plates makes car batteries very heavy, especially because there is more than just one pair of plates. One cell consists of one pair of lead and lead dioxide plates. The half-reactions and their standard potentials are found in Table 11.1:

$$PbO_2(s) + 4\ H^+(aq) + SO_4^{2-}(aq) + 2e \rightarrow PbSO_4(s) + 2\ H_2O + 1.69\ v$$

$$Pb\ (s) + SO_4^{2-}(aq) \rightarrow PbSO_4\ (s) + 2e \qquad\qquad +0.36\ v.$$

Together, the overall cell reaction in a lead storage battery is

$$Pb\ (s) + PbO_2(s) + 4\ H^+(aq) + 2\ SO_4^{2-}(aq) \rightarrow$$

$$PbSO_4\ (s) + PbSO_4(s) + 2\ H_2O.$$

The standard potential for a single cell is +1.69 + 0.36 = +2.05 v. There are six pairs of plates, so that the battery generates 12 v.

The high current makes working with car batteries more hazardous. Current is what kills, not voltage. Therefore, more caution has to be used with car batteries. Another difference is that current flow is what drains batteries, so high currents are draining car batteries quickly. Fortunately, current is drawn only briefly to start the engine. As soon as the engine begins running, some power from the engine itself causes the alternator (or generator) to begin recharging the battery. Recharging the battery is an example of an electrolytic process and will be discussed in the next section.

Important Point

Each cell in a car battery generates about 2 volts. Since six cells are connected in series, 6 x 2 = 12 v overall. Motorcycle batteries are smaller. They only have three cells, so they generate 6 v.

A car's accessories also draw current. Most of us know from experience that leaving the headlights on (or even the interior lights, if left on long enough), drains the battery sufficiently that it cannot start the car anymore. We call it a dead battery, but it is not really dead—it just needs to be recharged.

Electrolytic Cells

Not all electrochemical reactions take place spontaneously, that is, without the input of external energy to make them take place. Oxidation-reduction reactions that do not occur all by themselves are made to take place in an electrolytic cell by supplying energy in the form of an electrical current. Current is supplied by an external direct current (DC) power supply. The cell has two electrodes: the anode, which is positively charged and is the site at which the oxidation half-reaction takes place; and the cathode, which is negatively charged and is the site at which reduction occurs.

In an electrolytic cell the positive electrode of the cell is connected to the positive electrode of the power supply, and the negative electrode is connected to the negative electrode of the power supply. The definitions of anode and cathode are the same as for a galvanic cell, but the polarities have been reversed.

Important Point

The opposite of any spontaneous process is a nonspontaneous process. Spontaneous oxidation-reduction reactions can take place in galvanic cells. Their opposites require electrolytic cells to take place.

Industry makes extensive use of electrolytic cells. Aluminum metal is recovered from aluminum ore by the Hall-Héroult process, named for its inventors, American chemist Charles Martin Hall (1863–1914) and French chemist Paul Héroult (1863–1914). Sodium metal and chlorine gas are recovered from seawater by electrolysis. Electrolysis is often used to obtain samples of metals, e.g. copper, in very pure form.

Historical Note

Charles Hall and Paul Héroult both were born and died in the same years. Living on opposite sides of the Atlantic Ocean, they independently developed the electrolytic processes for obtaining aluminum metal from its ores. In the United States, Hall founded an aluminum company that eventually became ALCOA, the Aluminum Company of America.

When you plug batteries or any electronic device—cell phone, iPod, iPad, laptop computer, etc.—into a recharging unit, electrolysis is occurring. When you start your car, the battery supplies energy—it acts as a galvanic cell, but only briefly. After that, as the combustion of gasoline supplies the energy to keep the motor running, energy from the motor also acts to recharge the battery through the car's alternator.

Conclusion

Electrochemical reactions are taking place all around us. Most of us own electronic devices that run on batteries, or at least flashlights that use batteries. When the batteries run down, either they have to be disposed of or recharged. Any use of batteries as a source of energy is an example of galvanic action. Anytime batteries are being recharged is an example of electrolysis.

CHAPTER 12

Energy and Chemical Reactions

In This Chapter

➤ History of energy

➤ Energy and work

➤ Conservation of energy

Energy is a familiar concept, especially energy due to motion, electrical energy that powers our appliances and lightbulbs, and chemical energy that is released when fireworks light the night sky on the Fourth of July. There are many different forms of energy. These forms can convert into each other. As they do so, the chemical and physical changes that take place around us are governed by general principles that are called the laws of thermodynamics.

Chem Vocab

Thermodynamics is the branch of science concerned with the effects of work, heat, and other forms of energy on a system.

Much of chemistry deals with the properties of matter at the microscopic level of atoms and molecules. We live, however, in a world that is much larger than atoms and molecules. We live in a macroscopic world in which we experience matter in bulk quantities that we can see, touch, taste, and smell. Thermodynamics describes just such a world. The word thermodynamics means "heat" and "work," or the conversions of heat energy into work and vice versa.

Thermodynamics is one of the three pillars of classical physics (physics before 1900); the other two pillars are mechanics and electromagnetism. In chemistry, thermodynamics looks into the amount of heat energy given off or absorbed during chemical reactions.

A Brief History of Energy

In the beginnings of human history, the only source of energy people had with which to do work was their own strength. The ability to harness and control fire added a new form of energy—heat—although it was tens of thousands of years before scientists actually recognized heat as a form of energy. Animals were domesticated, which meant that human muscle could be supplemented by animal muscle. In developing countries of the world today, which includes billions of people, human and animal muscle are still largely the main sources of energy by which work gets done, and the primary fuels are still wood, charcoal, plant waste and manure.

But in industrialized countries, machines have been invented that do much of our work for us. Energy for cooking, space heating, and transportation comes not from raw materials gathered by hand from the environment around us, but mostly from fossil fuels, with only small contributions from other sources of energy—hydroelectric, nuclear, wind, solar, and tidal.

In Sir Isaac Newton's time (1642–1726) and during the following century, science consisted mostly of the study of motion as governed by Newton's laws of motion, a field still known today as mechanics. In the 1700s, an understanding of mechanical processes spawned the industrial revolution, marked importantly by the invention of the steam engine.

The steam engine was truly an astounding machine. Initially using heat from the combustion of wood, and later from the combustion of coal, machines could do work and convey people across the land without an expenditure of human or animal muscle. Steam boats were invented, along with locomotives, factory machinery, electrical power plants, and in the twentieth century, automobiles, airplanes, and spaceships.

Historical Note

The science of thermodynamics arose at the beginning of the industrial revolution with attempts to make steam engines operate more efficiently.

Unknown to people of the eighteenth and nineteenth centuries, a complex series of energy transformations had been accomplished. Energy produced by nuclear processes in the core of the Sun travels to Earth in the form of light, where the energy is fixed in green plants through the process of photosynthesis. Mature trees eventually are cut down, or fossil fuels are extracted from the ground. Either way, carbon-based fuel is burned, producing heat, which—in a steam engine—boils water. The boiling water becomes steam, which spins turbine blades, turning heat energy into mechanical energy. Those spinning turbine blades can turn propellers on ocean liners or spin electromagnets in power plants and create electricity. These same steam engines, and their modern descendants—gasoline, diesel, and jet turbine engines—can power boats along our waterways, propel railroad trains along their tracks, send trucks and automobiles down our interstate highways, and lift heavy aircraft tens of thousands of feet into the atmosphere.

Energy and Work

Since thermodynamics involves the conversion of energy from one form into another, it is important to understand what different kinds of energy there are in the world. The kinds of energy in the following discussion are the ones that are most often the subject matter of physics and chemistry.

Kinetic Energy

Kinetic energy is the energy of motion. Any object that is moving is said to possess kinetic energy. The equation used to calculate kinetic energy is

$$KE = \frac{1}{2}mv^2,$$

where m = mass, and v = speed. If mass is expressed in kilograms (kg) and speed in meters per second (m/s), then kinetic energy is expressed in $kg\text{-}m^2/s^2$, which is abbreviated as the joule, after the English physicist James Prescott Joule (1818–89). The symbol for the joule is J, and 1,000 joules are equal to 1 kilojoule (kJ).

Gravitational Potential Energy

The energy due to position in a gravitational field is called gravitational potential energy. Taking the ground or the floor of a room as the reference level, any object that is higher than the ground or the floor is said to possess gravitational potential energy. An object that is lifted to a short height, h, above the ground has a potential energy given by the equation

PE = mgh, where m = mass, g = acceleration due to gravity (9.8 m/s^2, or 32 ft/s^2), and h = height. If height is measured in meters (m), potential energy is expressed in units of joules. Note that the weight (W) of the object is equal to the product mg, so we could also write that PE = Wh.

Chem Vocab

Together, kinetic and gravitational potential energy are referred to as mechanical energy.

Thermal Energy

The heat content of an object is called thermal energy and is related to the object's temperature. Hot objects have a high heat content; cold objects have a lower heat content. Before heat was recognized as a form of energy, it had its own unit, the calorie (cal). The calorie was defined as the amount of heat required to raise the temperature of 1 gram of water 1°C (or, more precisely, to raise the temperature of 1 gram of water from 14.5°C to 15.5°C). Today, physicists and chemists almost always use joules, and the calorie is now defined as being equal to 4.184 joules. The quantity of any kind of energy nowadays is usually expressed in units of joules.

Chemical Energy

Chemical energy is the energy released when a bond is formed between two atoms, as well as the energy that must be absorbed to break a bond. Chemical energy is converted into heat during the combustion of fossil fuels.

Radiant Energy

The energy associated with light or other forms of electromagnetic radiation is called radiant energy. High-energy forms of radiation are ultraviolet light, X-rays, and gamma rays. Low-energy forms of radiation are radio waves, microwaves, and infrared radiation. Visible light represents a very, very narrow range of energy that falls in-between high- and low-energy forms of radiation.

Electrical Energy

The energy associated with an electrical current is called electrical energy. Current flow is due to the motion of electrons. As with kinetic energy, there is motion. What is important in an electrical current, however, is not the fact that mass is moving, but that charge is moving. Current flows because of an electrical potential difference between two parts of a circuit, such as between the positive and negative electrodes of a battery. Current (I) is measured in units of amperes (A), named after the Italian physicist Andrè Marie Ampère (1775–1836). Potential difference (V, or EMF for "electromotive force") is measured in units of volts (also V), named after the Italian physicist Alessandro Volta (1745—1827), who was also the inventor of the first battery. The energy associated with an electrical current is given by

$E = VIt$. (Thus, the product of volts times amperes times time is equivalent to joules.)

Important Point

An object's mechanical energy varies with the mass of the object and is independent of whether or not the object has an electrical charge. An object's electrical energy varies with the electrical charge on the object and is independent of the mass of the object.

Nuclear Energy

Nuclear energy is the energy associated with mass. Before the German physicist Albert Einstein (1879–1955), everyone assumed that mass and energy were two completely different quantities—mass being associated with material objects, whereas energy has no material substance to it.

One of Einstein's great contributions to physics, however, was showing that mass and energy are equivalent, as expressed in possibly the most famous formula in all of science: $E = mc^2$, where m = mass and c = the speed of light (3.00×10^8 m/s, or 186,000 miles/s). Now c is a very big number, so c-squared is a really big number! A very tiny change of mass results in a huge amount of energy. The equivalence of mass and energy says that any time the energy of an object changes, its mass also changes.

This means that a car moving down the highway has more mass than the same car at rest, that a book lifted 2 feet above a desktop has more mass than the same book when sitting on the desk, and that the products of a chemical reaction really do not have exactly the same total mass as the reactants, since at least some heat is given off or absorbed by every chemical reaction. These changes in mass, however, are so miniscule that we neglect them. In nuclear reactions, on the other hand, since an enormous amount of energy is released, the changes in mass are appreciable.

Work

Since thermodynamics is the science of the relationship between heat and work, it is important to define *work*. Doing work on an object changes the energy of the object. Mechanical work changes either the object's kinetic energy by making the object speed up or slow down, or the object's gravitational potential energy by lifting or lowering the object. Sometimes the work being done changes both an object's kinetic and potential energy. Such is the case, for example, when a jet's engines lift the jet off the ground and accelerate the jet at the same time. Since work = change of energy, work also has units of joules.

In many situations, work (w) is thought of as the product of force (F) times distance (d), or $w = F \cdot d$. If force is in pounds (lb) and distance is in feet (ft), then work is expressed in foot-pounds (ft·lb). If force is in newtons (N) and distance is in meters (m), then work is expressed in newton·meters (N·m), which is the same as joules. In the case of the expansion or contraction of gases, if pressure (P) is constant, then work is the product of pressure times change of volume (V), i.e., $w = P\Delta V$. If pressure is in pascals (pa, which is the same as newtons per meter^2) and volume is in cubic meters (m^3), then work is again expressed in newton·meters, or joules. (In situations where the pressure is not held constant, the calculation is more complicated.)

Special Kinds of Processes

Scientists have special vocabulary terms to describe processes that take place under prescribed sets of conditions. Some of these terms are the following:

➤ Isobaric: A term that means "constant pressure." For an isobaric process, $w = P\Delta V$. This is typically the case in chemical reactions that occur in vessels that are open to the atmosphere. If gases form during the reaction, the gases can expand into the surrounding atmosphere.

➤ Isochoric: Any process in which the volume does not change. Since $\Delta V = 0$, then $w = 0$ also. Processes that take place in containers with rigid walls are isochoric.

➤ Isothermal: Any process that occurs at constant temperature. The vessels in which chemical reactions take place may be thermostatically controlled so that the temperature cannot change.

The Law of Conservation of Energy

The law of conservation of energy is also called the first law of thermodynamics. Amazingly, as monumental as the contributions of Sir Isaac Newton were to physics in the latter half of the seventeenth century, Newton did all of his work ignorant of conservation of energy because he did not recognize heat as being a form of energy. In the 1780s, the French chemist Antoine Lavoisier stated the law of conservation of mass, but he did not recognize heat as a form of energy either. In fact, Lavoisier included heat (or the caloric as he called it) in his list of elements. It was not until the middle of the nineteenth century that James

Joule recognized heat as a form of energy and established what scientists call the mechanical equivalence of heat.

The mechanical equivalence of heat says that any change in the energy of a system that can be accomplished by doing mechanical work on the system can also be accomplished by adding heat to the system. In fact, we would say that energy is conserved in all mechanical processes precisely because heat is a form of energy. A rolling ball that eventually comes to rest loses its kinetic energy, but not its total energy; because of friction, the ball's kinetic energy is converted to heat, but heat is still a form of energy. Joule, therefore, is credited with the discovery of the first law of thermodynamics, which can be stated in various forms, but fundamentally is what scientists call the law of conservation of energy.

Historical Note

Lavoisier stated the law of conservation of mass in the late 1700s. Joule formulated the law of conservation of energy in the mid-1800s. Half a century later, in 1905, as part of his special theory of relativity, Einstein stated that mass and energy are equivalent. Today, the two laws are given most precisely as the combined law of conservation of mass–energy.

To use the first law, we define a system of interest such as a car engine, or a book on a desk, or a beaker in which a chemical reaction is taking place, or a cell inside a tissue of a living organism, or other examples that may be of physical, chemical, or biological interest. Anything that is not part of the system is part of the surroundings. We can add energy from the surroundings to the system, or the system can give off energy to its surroundings.

In this chapter we are interested only in mechanical energy in the form of work and thermal energy in the form of heat, but the same principles would apply for other forms of energy such as electrical. Mathematically, we state that $\Delta E_{system} = q + w$, where E is the internal energy of the system, q is the heat the system absorbs from its surroundings, and w is the work that something in the surroundings does on the system. Note the following sign conventions:

Heat, q	Work, w
q > 0 if the system absorbs heat.	w > 0 if the surroundings do the work.
q < 0 if the system gives off heat.	w < 0 if the system does the work.
q = 0 if no heat is exchanged.	w = 0 if no work is done.

Important Point

It is rather emphatic to understand that a process that is adiabatic is not isothermal. In an isothermal process, the temperature does not change precisely because heat is exchanged between the system and the surroundings. In an adiabatic process, however, heat cannot be exchanged, so the temperature of the system must change.

Once again, chemists have a special vocabulary to describe the different kinds of heat exchanges:

➤ If the system absorbs heat from its surroundings, the process is said to be endothermic.

➤ If the system gives off heat to its surroundings, the process is said to be exothermic.

➤ If no heat is being exchanged between the system and its surroundings, then the system must be thermally insulated from its surroundings, and the process is said to be adiabatic.

Important Point

There are various ways to state the first law of thermodynamics:

➤ The energy of the universe is conserved.

➤ The energy of the universe is a constant.

➤ Energy can be converted among its various forms, but energy can never be created nor destroyed.

➤ The energy of a closed system is constant.

➤ The change in energy of a system is equal to the heat absorbed by the system from its surroundings plus the work the surroundings do on the system.

Examples of Different Kinds of Work and Heat Exchanges

Interconversions of heat and work occur all around us. Here are some examples.

Automobile Engines

In terms of its basic features, the fuel/air mixture in a cylinder of an automobile engine is a good representative example of a system that can do work or that has work done on it by a moving piston. When the valve opens, a mixture of gasoline vapors and air (which contains oxygen) enters the cylinder. The piston then moves in a direction so as to compress the gas. During that compression stroke the piston (the surroundings) is doing work on the fuel/air mixture (the system). Since the surroundings are doing the work on the system, work is positive. Because air does not conduct heat very well, and the compression stroke is of very brief duration, there isn't time for heat to be exchanged and the stroke is effectively adiabatic ($q = 0$). Therefore, $\Delta E = q + w = 0 + w > 0$.

When a force compresses a gas, or a gas expands against a restraining force (such as a piston), the change of energy of the gas is given by $\Delta E = C \cdot \Delta T$, where C is the gas's heat capacity. During the compression stroke, the energy of the fuel/air mixture increases, which also means the temperature increases.

In the case of a diesel engine, the heat of compression is sufficient to ignite the fuel/air mixture and the diesel fuel vapors burn. In the case of a gasoline engine, however, when the fuel/air is fully compressed, the spark plug ignites the gas mixture. Either way, the volume of the exhaust gas mixture is greater than the volume of the fuel/air mixture, so the exhaust gases expand. In the process, the expanding gases (which are still the system in this example) are now pushing against the piston (the surroundings). This means that the gases are doing work on their surroundings and W is now negative. Because the expansion happens so quickly, the process is again effectively adiabatic, so $q = 0$ and $\Delta E = 0 + w < 0$. Because the exhaust gases are losing energy, the gases cool down.

Another valve opens, removes the exhaust gas mixture, and the cycle continues. Note that even though on a very short time scale each stroke is adiabatic, over time the engine does become hot. The heat that results from combustion does flow to the other parts of the engine.

Atmospheric Temperature Lapse Rate

In the atmosphere, warm air parcels tend to occupy a greater volume than surrounding cooler air parcels. Because the warm air parcels are less dense than their immediate surroundings, buoyancy causes them to rise relative to their surroundings. The air parcels rise until they reach a level of the atmosphere that has a lower density than they do. At that level, the warm air expands against the surrounding air, doing work on the surroundings in the process.

Again, air is not a very good conductor of heat, so the expansion is effectively adiabatic. This means that q = 0, w < 0, and ΔE < 0. The net result is that as the energy of the rising air parcel decreases. Since ΔE still equals C·ΔT, ΔT < 0 and the air parcel cools off. The result is that the atmosphere cools off with increasing altitude above Earth's surface.

Chem Vocab

The atmosphere's adiabatic lapse rate measures the rate at which a parcel of air cools as it rises in the atmosphere. In dry air, the lapse rate is 9.8°C for an increase in altitude of 1,000 meters, which is equivalent to 5.4°F every 1,000 feet.

In dry air, this cooling-off effect is called the adiabatic lapse rate of the atmosphere. Anyone who has hiked up or driven to the top of a mountain has experienced the effect of the air cooling off during the ascent.

Phase Changes

Consider the following sequence of phase changes: solid → liquid → vapor. Solids must absorb heat in order to melt, and liquids must absorb heat in order to vaporize. Both are endothermic processes. When a solid melts, there isn't very much change in volume (maybe 10 percent or so), so the work done when the solid expands is usually ignored. (Of course, in the case of water, the solid contracts as it melts—more about that in Chapter 24.) On the other hand, there is a big change in volume when a liquid vaporizes—the increase in volume is on the order of a factor of 1,000.

A liquid must do work of expansion against the surrounding atmosphere as the liquid vaporizes and the vapor expands. Consequently, a liquid needs a lot of energy in order to vaporize. A liquid must absorb much more heat to convert to the vapor phase than is required for a solid to convert to the liquid phase. Ice, for example, requires 333 joules of heat per gram to melt at 0°C, in contrast to liquid water, which requires 2,257 joules per gram to vaporize at 100°C. The first number is called the heat of fusion of water, and the second number is called the heat of vaporization of water. Different substances will have different values for their heat of fusion and vaporization.

The reverse sequence of phase changes—vapor → liquid → solid—are all exothermic. In each case, heat must be given off. Gases must give off heat in order to condense into the liquid phase, and liquids must give off heat in order to solidify or crystallize. This brings up an important point to remember: The reverse of any endothermic process is an exothermic process, and vice versa.

Chem Vocab

Heat of fusion is the amount of heat required to melt a solid.

Heat of vaporization is the amount of heat required to vaporize a liquid.

Conclusion

The law of conservation of energy (or of mass-energy) is one of the fundamental laws of the universe. We often hear about the "production" of energy. Energy cannot be produced (created), however. The term *production* just refers to converting energy from one form into another form, such as converting the chemical energy stored in coal into heat and ultimately into electrical energy.

We also often hear about nonrenewable energy sources versus renewable sources. Fossil fuels are nonrenewable because once consumed, they are gone. Ethanol is renewable because it comes from grains like corn, and we can plant more corn every year. Solar and wind are renewable because they both depend on energy from the Sun, which we expect will continue to shine for at least another five billion years.

Whether the end uses of energy are for electricity, space heating, or transportation, in most cases energy conversions must take place. Energy conversions tend to be expensive, and in many cases are not very efficient, which is why developed countries consume such an enormous amount of energy. As world population continues to increase, it makes sense to conserve energy as much as possible. Although in coming decades we may expect to see renewable energies such as solar and wind playing ever increasing roles, conservation should be playing an equally increasing role.

CHAPTER 13

Radioactive Processes and Applications

In This Chapter

➤ Symbols used in nuclear reactions
➤ The discovery of radioactive decay
➤ The nature of radioactive decay
➤ Half-Life
➤ Applications

During the Middle Ages, alchemists tried to transmute, or change, lead into gold. They treated lead with all sorts of different chemical reagents, with absolutely no success. Alchemists even kept their notes in their own secret codes, so that if they did succeed in making gold, no one would be able to steal their secret. The reason they never succeeded, however, is because transmutation is not an ordinary chemical reaction—a change in the arrangement of atoms—but rather it is a nuclear reaction—a change in the number of protons and neutrons in the nuclei of atoms.

Chem Vocab

To transmute a substance means to change it into a different substance. In chemistry, transmutation usually refers to changing an element into a different element.

It was not until the dawn of the twentieth century that the nucleus was even discovered or that chemists discovered how to change protons into neutrons or vice versa. Thus, nuclear chemistry began relatively recently in history compared to the everyday sorts of chemistry that people had been doing since the Stone Age.

Symbols Used in Nuclear Reactions

Chemists usually study four fundamental kinds of nuclear reactions:

➤ Radioactive decay

➤ Transmutation of elements

➤ Nuclear fission

➤ Nuclear fusion

Chem Vocab

A nuclide is a specific nuclear species characterized by its proton number *Z* and its neutron number *N*.

Isotopes are nuclides with the same number of protons but a different number of neutrons.

In each case, the electrons in the atoms are ignored. All of the symbols used to write equations for nuclear reactions represent only the atomic nuclei. Recall that we said we write the symbol for an isotope in the form $^{A}_{Z}X$, where X is the symbol for the element, Z is the element's atomic number (which is determined by the number of protons in atoms of that element), and A is the element's mass number (the sum of the number of protons and neutrons in atoms of that isotope). If we are concerned only with an atom's nucleus, we may also use the term *nuclide* instead of *isotope*, since the term *isotope* commonly refers to the entire atom and includes the atom's electrons.

Other Particles besides Atomic Nuclei

Besides atomic nuclei themselves, there may be other elementary particles that are reactants or products in nuclear reactions. It is important to recognize their symbols. These particles are shown in Table 13.1.

Table 13.1: Elementary Particles

Particle	Symbol	Charge	Mass Number
Proton	$^{1}_{1}p$	+1	+1
Neutron	$^{1}_{0}n$	0	+1
Electron	$^{0}_{-1}e$, or β^{-}	−1	0
*Positron	$^{0}_{+1}e$, or β^{+}	+1	0
Photon	γ	0	0

*A positron is also called an *antielectron* and is an example of *antimatter*. An *antiparticle* has the same mass but the opposite charge of an ordinary particle.

The Discovery of Radioactive Decay

Radioactive decay was the first kind of nuclear transformation to be discovered. In 1896, the German physicist Wilhelm Röentgen (1845–1923) published his discovery of X-rays, which are a very high-energy form of light, higher in energy than ultraviolet light, and invisible to humans. For this discovery, Röentgen was awarded the first Nobel Prize in physics (in 1901) and was commemorated in 2004 in the periodic table by having element 111 named after him—röentgenium (Rg).

Following Röentgen's announcement of X-rays, the French physicist Antoine Henri Becquerel (1852–1908) began studying X-rays. Becquerel's experiments led him serendipitously to the discovery of radioactivity in samples of uranium salts in his possession. For this discovery, Becquerel shared the 1903 Nobel Prize in physics with his French colleagues physicist Pierre Curie (1859–1906) and Pierre's wife, Polish-born chemist Marie Sklodovska Curie (1867–1934), who both had made significant contributions to an early understanding of what is involved when radioactive decay occurs.

In the early days, their work was very difficult. It was especially hampered by the fact that the electron was not even discovered until 1897, the proton shortly thereafter, and the neutron not until 1932. Even the existence of the atomic nucleus was unknown until its discovery in England in 1911 by the New Zealand–born physicist Ernest Rutherford (1871–1937). Scientists also had not yet understood the concepts of atomic number or isotopes.

Important Point

Every elementary particle has an antiparticle—a particle that has the same mass as the ordinary particle but the opposite charge.

Historical Note

Röentgen was not looking for X-rays and Becquerel was not looking for radioactivity. Both discoveries were completely serendipitous.

Historical Note

Pierre Curie died prematurely in a tragic accident. Marie Curie carried on their work. Her research resulted in the discovery of the radioactive element radium, an accomplishment for which she was awarded a second Nobel Prize, this time in chemistry, in 1911.

The Nature of Radioactive Decay

There are several modes by which radioactive decay can occur. The characteristics of the three most common modes were elucidated in 1903 by Rutherford and Frederick Soddy (1877–1956) and named alpha (α), beta (β), and gamma (γ) decay after the first three letters of the Greek alphabet.

Historical Note

Rutherford has been called the father of nuclear physics. His lifetime of accomplishments earned him the Nobel Prize in chemistry (not physics) in 1908, knighthood in 1914, the Order of Merit in 1921, and a place in the periodic table—element 104, rutherfordium (Rf). Soddy, an English chemist, received his own Noble Prize in 1921 for developing the concept of isotopes.

Alpha Decay

An alpha particle (α) is a helium ion. Specifically, it is the +2 ion of the helium isotope $_2^4He$. Consisting of a bare nucleus that contains two protons and two neutrons, an alpha particle has a charge of +2 and a mass number of +4.

Important Point

In alpha decay, the atomic number of the element decreases by two and the mass number decreases by four. Because its atomic number changes, the identity of the element changes.

Alpha decay tends to occur mostly among the heavy elements in the periodic table, elements with atomic numbers greater than 82 (lead). Because the number of protons in the parent isotope decreases by two, the atomic number of the daughter isotope (the decay product) is two units less than the atomic number of the parent isotope, and the daughter corresponds to an element that lies two squares to the left of the position occupied by the parent isotope. Because the number of neutrons also decreases by two, the mass number of the daughter isotope is four units less than the mass number of the parent.

Here are several examples of alpha decay:

➤ $^{238}_{92}\text{U} \rightarrow {}^{234}_{90}\text{Th} + {}^{4}_{2}\text{He}$

➤ $^{226}_{88}\text{Ra} \rightarrow {}^{222}_{86}\text{Rn} + {}^{4}_{2}\text{He}$

➤ $^{210}_{84}\text{Po} \rightarrow {}^{206}_{82}\text{Pb} + {}^{4}_{2}\text{He}$

These equations illustrate two important conservation laws that must be obeyed when writing equations for nuclear reactions. Just as we do in an ordinary chemical reaction, we balance the number of atoms of each element on both sides of the equation. In a nuclear reaction, we also balance subscripts and superscripts on both sides of the equation. (Unlike the practice in an ordinary reaction, we do not concern ourselves with balancing charges in a nuclear reaction since in a nuclear reaction we are ignoring the electrons anyway and only indicating the nuclei of the atoms.)

If the species in the reaction are nuclides of elements, the subscripts represent the elements' atomic numbers, or, equivalently, the charges on their nuclei. A fundamental physical law is the law of conservation of charge, which must be obeyed in all chemical and physical processes. Conservation of charge requires that the sum of the charges on the right side of the equation must equal the sum of the charges on the left side. That means that the sum of the subscripts on the right side must equal the sum of the subscripts on the left side. In the equations above, we see that $92 = 90 + 2$; $88 = 86 + 2$; and $84 = 82 + 2$.

Important Point

The law of conservation of charge requires that the sum of the subscripts on the right side of a nuclear equation must equal the sum of the subscripts on the left side of the equation.

The superscripts equal the number of protons plus neutrons. Protons and neutrons are examples of a family of elementary particles called baryons (from the Greek word *barus*, which means "heavy"). Just as charge is conserved, the number of baryons must also be conserved. This means that the sum of the superscripts on the right side of the equation must equal the sum of the superscripts on the left side. In the three equations shown above, we see that $238 = 234 + 4$; $226 = 222 + 4$; and $210 = 206 + 4$.

Important Point

Conservation of baryon numbers requires that the sum of the superscripts on the right side of a nuclear equation must equal the sum of the superscripts on the left side of the equation.

Important Point

In beta decay, the atomic number of the element increases by one, but the mass number does not change. Because its atomic number changes, the identity of the element changes.

Beta Decay

A beta particle (β) is an electron. It has a charge of −1, a mass number of 0, and is symbolized as $_{-1}^{0}e$. Beta decay can occur among elements anywhere in the periodic table, from hydrogen to the very heaviest elements. During beta decay a neutron inside the atom's nucleus is converted into a proton. Therefore, the number of protons in the nucleus increases by one unit. This means that the atomic number of the daughter isotope is one unit more than the atomic number of the parent isotope, and the daughter corresponds to an element that lies one square to the right of the position occupied by the parent isotope. The mass number, however, does not change because the sum of protons and neutrons remains constant.

Here are several examples of beta decay:

➤ $_{1}^{3}H \rightarrow {}_{2}^{3}He + {}_{-1}^{0}e$

➤ $_{27}^{60}Co \rightarrow {}_{28}^{60}Ni + {}_{-1}^{0}e$

➤ $_{93}^{239}Np \rightarrow {}_{94}^{239}Pu + {}_{-1}^{0}e$

Note the same conservation laws that we observed for alpha decay: the sum of the subscripts is the same on each side, and the sum of the superscripts is the same on each side.

Important Point

Because a gamma ray is a form of light, it has no mass or charge. Because its charge does not change, an atom that undergoes gamma decay does not change its identity.

Gamma Decay

A gamma ray (γ) is a high-energy form of light. Since light carries neither mass nor charge, the identity of the nuclide does not change during gamma decay. Gamma decay follows other nuclear processes. It occurs to lower the energy of the daughter nucleus to make it more stable (we say that the nucleus relaxes) but otherwise does not change the nucleus. An example is the decay of Th–234 after its formation from U–238, as shown:

➤ $_{92}^{238}U \rightarrow {}_{90}^{234}Th + \alpha$

➤ $_{90}^{234}Th \rightarrow {}_{90}^{234}Th + \gamma$

Other Modes of Radioactive Decay

Two other modes of radioactive decay are positron decay and electron capture. A positron is an electron with a positive charge and is symbolized $^{0}_{+1}e$ or β^+. Positron decay is, in a sense, the opposite of beta decay—instead of a neutron turning into a proton, a proton turns into a neutron. The result is to decrease the atomic number by one unit, again without changing the mass number. This time, therefore, the daughter isotope is located one space to the left of the parent, instead of one space to the right.

Positron decay is much less common than beta decay but also can occur throughout the periodic table. Here are several examples of positron decay:

➤ $^{11}_{6}C \rightarrow {}^{11}_{5}B + {}^{0}_{+1}e$

➤ $^{18}_{9}F \rightarrow {}^{18}_{8}O + {}^{0}_{+1}e$

➤ $^{65}_{30}Zn \rightarrow {}^{65}_{29}Cu + {}^{0}_{+1}e$

Electron capture (EC) is also less common than beta decay, but it has the same net result as positron decay—the daughter isotope is located one space to the left of the parent. In electron capture, an orbiting electron is taken into the nucleus where it combines with a proton to form a neutron. An example of electron capture is the decay of silver–108, as shown:

$$^{108}_{47}Ag + {}^{0}_{-1}e \rightarrow {}^{108}_{46}Pd.$$

Half-Life

Each radioactive nuclide decays at its own characteristic rate. That rate is completely independent of the temperature, pressure, or volume of the sample, or whether the sample is a pure element, part of a compound, or part of a mixture. The most convenient way to distinguish the decay rate of different nuclides is to state their half-lives. The half-life ($t_{1/2}$) of a radioactive substance is the time it takes for one half of the atoms that were present initially to decay. Half-lives can be very, very short—on the order of microseconds (10^{-6} s, or μs)—or very, very long—on the order of billions of years.

If one half of the atoms initially present decay during the first half-life, then one half of the initial atoms are still left. Then, one half of those atoms will decay during the second half-life, so that one-quarter of the initial atoms are still left. One half of those will decay during the third half-life, leaving one-eighth of the initial atoms remaining, and so forth.

Table 13.2: Fraction of Initial Sample of a Radioactive Isotope Remaining after N half-lives

Number of Half-Lives, N	Fraction Remaining, f
0	1.00 = 1
1	0.50 = 1/2
2	0.25 = 1/4
3	0.125 = 1/8
4	0.0625 = 1/16
5	0.03125 = 1/32
6	0.015625 = 1/64

Table 13.3: Half-Lives of Selected Radioactive Isotopes

Isotope	Decay Mode	Half-Life
H–3	β	12.26 y
C–14	β	5730 y
K–40	EC	1.28 by
Ge–68	EC	275 d
Sr–90	β	28.8 h
Mo–99	β	66.02 h
I–131	β	8.040 d
Pm–147	β	2.62 y
Po–210	α	138.38 d
Rn–222	α	3.824 d
Ra–226	α	1,600 y
Th–232	α	1.40 by
U–238	α	4.47 by
Lr–260	α	3.0 m
Sg–263	α	0.9 s

Key to Units: s = second, m = minute, h = hour, d = day, y = year, by = billion years

The Equation for Radioactive Decay

Let N = number of radioactive atoms currently present, N_0 = initial number of radioactive atoms, $t_{1/2}$ = half-life of the isotope, and t = elapsed time. Then the equation used is the following:

$$N = N_0 e^{-0.693t/t_{1/2}},$$

where e is the base of natural logarithms.

In logarithmic form, this equation becomes

$$\ln(N/N_0) = -0.693t/t_{1/2}.$$

Depending upon what variables are already known, these equations can be solved for N, N_0, t, or $t_{1/2}$.

Important Point

In the equation for radioactive decay, *N* can stand for any unit of quantity—numbers or moles of atoms, or numbers of grams.

Applications of Radionuclides in Science

An important application of radioactivity in science is its use to determine the ages of human artifacts and of rocks and sediments. Dating techniques rely on naturally occurring radionuclides. Archaeologists use carbon dating to determine the age of human artifacts. Geologists use long-lived radioactive materials such as uranium and thorium to determine the age of rocks and Earth.

Carbon Dating

Carbon–12 is a stable isotope and is by far the most abundant isotope of carbon found on Earth. Carbon–13 is also stable but occurs on Earth only as a very small proportion of the total carbon. Carbon is found in large quantities in the atmosphere (mostly as CO_2, the most abundant atmospheric gas after N_2, O_2, Ar, and water vapor), the lithosphere (especially in rocks, like limestone, that contain the carbonate ion (CO_3^{2-}), and the hydrosphere (as dissolved CO_2 and as H_2CO_3, HCO_3^-, and CO_3^{2-}). Carbon is constantly cycling through the air, water, and rocks on Earth, as well as through living organisms, since carbon is the basis of life as we know it.

Carbon dioxide is removed from the atmosphere partly by its solution in lakes and oceans, but mostly through the process of photosynthesis in green plants, as shown:

$$6\ CO_2\ (g) + 6\ H_2O\ (l) \rightarrow C_6H_{12}O_6\ (s) + 6\ O_2\ (g).$$

Photosynthesis requires the presence of chlorophyll in green plants and energy from sunlight. The products are glucose ($C_6H_{12}O_6$) and oxygen gas (O_2). Glucose becomes the starches and cellulose that make up plants. At the same time, this is the source of oxygen gas in the atmosphere that is vital for both animals and plants.

Most of the carbon taken in by plants during photosynthesis is in the form of carbon–12, which is not radioactive. However, a small amount of radioactive carbon–14 (half-life = 5,730 y) is produced in the upper atmosphere as a result of cosmic rays interacting with atmospheric gases. The carbon–14 atoms that result are incorporated into atmospheric CO_2, which we can denote as $^{14}CO_2$. When $^{14}CO_2$ is taken in by plants during photosynthesis, the glucose that is produced contains carbon–14. During the lifetime of a plant—grass, herb, bush, or tree—the ratio $^{14}C/^{12}C$ becomes a relatively constant quantity.

Let us assume we are talking about a tree. The tree's ^{14}C is undergoing radioactive decay while the tree is still alive. However, most trees (with the exception of redwoods, bristlecone pines, creosotes, and a few other very long-lived species) live, at most, for only a few hundred years. Compared to 5,730 years, a few hundred years is a short enough period of time that very little ^{14}C decays during the lifetime of the tree.

When the tree dies—or is cut down for lumber, firewood, or some other use—photosynthesis ceases. At that moment, the ratio of ^{14}C to ^{12}C is the highest it will ever be. The fact that the tree is dead has no effect on the radioactive decay of the ^{14}C in it, so the ^{14}C continues to decay and the ratio of ^{14}C to ^{12}C decreases.

Suppose the tree had been cut down thousands of years ago so that the wood could be used in the construction of a house. Suppose that archaeologists today find that house during their excavation of a prehistoric site. They can cut a small sliver from the wood in the house and measure its rate of radioactive decay due to its ^{14}C content. In addition, the archaeologists could take a sliver from a living tree of the same species growing in the same area and measure its rate of radioactivity.

If the ratio of the rate of radioactive decay from the excavation site to the rate of decay from a fresh tree is 0.75, the archaeologists would use the equation for radioactive decay (above) to conclude that the tree that was used to build the house was chopped down about 2,400 years ago, giving the archaeologists an estimate of the age of the house, and perhaps a village of which the house was a part. (Ages are rounded off to allow for experimental error. A key assumption in the calculation is that the concentration of ^{14}C in the atmosphere has been constant since the tree died, which may or may not be strictly true.)

If the ratio of the two rates of decay is 0.50, the archaeologists would conclude that the house was one half-life in age, or 5,730 years. If the ratio is only 0.25, the archaeologists would conclude that the age of the house was two half-lives, or 11,500 years. This technique is considered to be reasonably reliable to about 10 half-lives, or roughly 57,000 years.

Carbon–14 dating has played an important role in archaeology since the discovery of C–14 in the 1930s.

Uranium and Thorium Dating

Uranium and thorium each have radioactive isotopes with half-lives of billions of years, or roughly the same order of magnitude as the age of Earth itself (see tables 13.4 and 13.5). These elements go through a series of radioactive decay that involve several radioactive isotopes with very different half-lives. The net result is the same, though. Both elements end their respective series with stable isotopes of lead. Both uranium and thorium are common enough in rocks that the ratios of uranium to lead or of thorium to lead in a rock can be used to date rocks that are billions of years old.

Table 13.4: Thorium–232 Decay Series

Parent Isotope	Daughter Isotope	Decay Mode	Half-life
$^{232}_{90}$Th	$^{228}_{88}$Ra	α	14 billion years
$^{228}_{88}$Ra	$^{228}_{89}$Ac	β	5.8 years
$^{228}_{89}$Ac	$^{228}_{90}$Th	β	6.2 hours
$^{228}_{90}$Th	$^{224}_{88}$Ra	α	1.9 years
$^{224}_{88}$Ra	$^{220}_{86}$Rn	α	3.7 days
$^{220}_{86}$Rn	$^{216}_{84}$Po	α	55.6 seconds
$^{216}_{84}$Po	$^{212}_{82}$Pb	α	0.15 second
$^{212}_{82}$Pb	$^{212}_{83}$Bi	β	10.6 hours
* $^{212}_{83}$Bi	$^{208}_{81}$Ti	α	1.1 hour
$^{208}_{81}$Tl	$^{208}_{82}$Pb	β	3.05 minutes
* $^{212}_{83}$Bi	$^{212}_{84}$Po	β	1.1 hour
$^{212}_{84}$Po	$^{208}_{82}$Pb	α	0.30 µs

* $^{212}_{83}$Bi **can decay by either alpha or beta decay.**

µs = microsecond

Table 13.5: Uranium–238 Decay Series

Parent Isotope	Daughter Isotope	Decay Mode	Half-life
$^{238}_{92}$U	$^{234}_{90}$Th	α	14 billion years
$^{234}_{90}$Th	$^{234}_{91}$Pa	β	24 days
$^{234}_{91}$Pa	$^{234}_{92}$U	β	0.84 second
$^{234}_{92}$U	$^{230}_{90}$Th	α	244 thousand years
$^{230}_{90}$Th	$^{226}_{88}$Ra	α	75 thousand years
$^{226}_{88}$Ra	$^{222}_{86}$Rn	α	1600 years
$^{222}_{86}$Rn	$^{218}_{84}$Po	α	3.8 days
* $^{218}_{84}$Po	$^{214}_{82}$Pb	β	3.0 minutes
$^{214}_{82}$Pb	$^{214}_{83}$Bi	α	27 minutes
* $^{214}_{83}$Bi	$^{210}_{81}$Tl	α	20 minutes
$^{210}_{81}$Tl	$^{210}_{82}$Pb	β	1.3 minutes
$^{210}_{82}$Pb	$^{210}_{83}$Bi	β	22.6 years
* $^{210}_{83}$Bi	$^{206}_{81}$Tl	α	3.0 million years
$^{206}_{81}$Tl	$^{206}_{82}$Pb	β	4.2 minutes
* $^{218}_{84}$Po	$^{218}_{85}$At	β	3.0 minutes
* $^{218}_{85}$At	$^{218}_{86}$Rn	β	1.6 seconds
$^{218}_{86}$Rn	$^{214}_{84}$Po	α	35 milliseconds
$^{214}_{84}$Po	$^{210}_{82}$Pb	α	164 microseconds
* $^{214}_{83}$Bi	$^{214}_{84}$Po	β	20 minutes
* $^{210}_{83}$Bi	$^{210}_{84}$Po	β	5.0 days
$^{210}_{84}$Po	$^{206}_{82}$Pb	α	138 days

* $^{218}_{84}$Po, $^{214}_{83}$Bi, $^{210}_{83}$Bi, $^{218}_{85}$At **can decay by either alpha or beta decay.**

Using these decay series, geologists have determined that the best estimate of the age of Earth is 4.6 billion years. The oldest rocks that have ever been found, however, are younger than this—3.5 to 4 billion years old. Because of plate tectonics—the movement of large masses of Earth's crust over Earth's mantle—rocks subside into Earth's interior, where they melt and become molten material. Later, the same material may again rise to Earth's surface or pour out as magma during a volcanic eruption. The rocks eventually cool, but when they do, they are different rocks than the ones they were before subsidence. Thus, no one has found a rock that is actually the same age as Earth itself.

Applications of Radionuclides in Medicine

In the area of human health, radioactivity is a double-edged sword. On one hand, radioactivity substances can make people sick and even kill them. On the other hand, radioactive substances can be used to diagnose and cure disease. Let's look first at the use of radionuclides in the diagnosis of disease.

Using Radioactive Substances to Diagnose Disease

Nuclear body scanning uses a small amount of radioactive materials to produce images that can distinguish between healthy and abnormal tissues. The images are similar to the ones produced by X-rays; the difference is that X-rays originate outside the body while radiation originates from sources that are injected into the body, swallowed, or inhaled.

One of the most common radionuclides medical technicians use is technetium–99m. (The *m* means that it is a metastable isotope. It is not completely stable, but it does have a somewhat longer half-life than would be expected.) Technetium is unusual in that it is one of only two elements lighter than lead (the other being promethium) that has no stable isotopes. Therefore, technetium has to be produced in nuclear reactors. Technetium is produced from molybdenum. Since Tc–99m has a half-life of only 6.0 hours, supplies must be produced continuously and then transported quickly to hospitals around the United States and Canada to be used before too much of it has decayed.

Technetium–99m is a gamma emitter. As the material passes through the organs of the body, a gamma ray camera records images. Abnormalities are apparent to a radiologist, a doctor who has had special training in analyzing such images. Technetium is excreted fairly readily in urine. That, and its short half-life, means that no special care needs to be taken after the procedure has been completed.

Lung scans are similar, except the radioactive material is inhaled. For liver scans, radioactive materials are injected into the body. As they pass through the liver, abnormal cysts or tumors will show up in the images.

Since iodine concentrates in the thyroid gland, a scan of the thyroid is conducted using a radioactive isotope of iodine, I–131, which has a half-life of 8.0 days. The iodine sample can be administered by swallowing a capsule or liquid. Only a small amount of radioactive iodine is used for the procedure, and after a week the level of radioactivity is negligible. Iodine–131 can also be administered to treat goiters.

Other radionuclides that are used in medical diagnosis include the following:

> ➤ Gallium–67: Half-life 62 hours; used to scan for infections or inflammations

> ➤ Indium–111: Half-life 7.5 days; used to scan for infections by taking a sample of the patient's blood, injecting the In–111, and then returning the blood to the bloodstream

> ➤ Thallium–201: Half-life 73 hours; used to scan the heart for evidence of damage to the heart muscle

Using Radiation Therapy to Treat Disease

Radiation therapy can also be used to treat disease. Probably the best-known example of radiation therapy is in the treatment of cancer. The strategy is to use radiation to selectively kill cancer cells. The radioactive isotope most commonly used is cobalt–60, which is a beta emitter with a half-life of 5.3 years. Radiation may be used to shrink a tumor prior to surgery, to kill any remaining cancer cells after surgery, or in combination with chemotherapy. A drawback of radiation therapy is that damage may also be done to the healthy tissue that surrounds a tumor.

Other radionuclides used in radiation therapy include cesium–131 (half-life of 9.7 days), which is used to fight prostate cancer, and phosphorus–32 (half-life of 14.3 days), which is used to treat leukemia.

Conclusion

Radioactive materials are all around us and can be very beneficial to us in our daily lives. Unfortunately, many people react in fear when they hear the word radioactive. An understanding of the nature of radioactive decay, however, and the uses of radioactive isotopes is something to be appreciated.

CHAPTER 14

Nuclear Reactions and Applications

In This Chapter

➤ Transmutation reactions

➤ Nuclear fission and fusion

➤ Synthesis of transuranium elements

➤ Production of superheavy elements

In Chapter 13 we learned about one kind of nuclear reaction—radioactivity. In this chapter we will learn about several other kinds of nuclear reactions and their importance in our society.

Transmutation Reactions

To transmute a substance is to change it into a different substance. Early alchemists attempted unsuccessfully to transmute lead into gold. The reason that they failed is because lead and gold are both elements. An element is identified by the number of protons found in its nucleus. The methods the alchemists used were ordinary chemical reactions that only involved interactions among the electrons in atoms, not changes to the nuclei of the atoms. To change one element into another element requires that the number of protons

Chem Vocab

Transmutation means change or transformation. The term is used to describe what happens when one element is changed into another element.

> ## Historical Note
>
> Today's chemists have succeeded in changing lead into gold. They have even changed platinum into gold. The costs are prohibitive, however, so there is no practical reason for doing either one.

be changed, and that takes a nuclear reaction.

Ernest Rutherford (1871–1937) was the British physicist who performed the first nuclear transmutation. In 1911, Rutherford bombarded nitrogen nuclei with alpha particles and succeeded in converting nitrogen into oxygen, as shown in the following equation:

$$^{14}_{7}\text{N} + {}^{4}_{2}\text{He} \rightarrow {}^{17}_{8}\text{O} + {}^{1}_{1}\text{p}.$$

Since that time nuclear reactors around the world have produced hundreds of isotopes of most of the elements in the periodic table. As discussed in Chapter 13, some of these isotopes are used daily in nuclear medicine. Technetium–99m is a gamma ray emitter used daily in hospitals. Tc–99m is produced from the beta decay of molybdenum–99 as shown in the following reaction:

$$^{99}_{42}\text{Mo} \rightarrow {}^{99m}_{43}\text{Tc} + {}^{0}_{-1}\text{e}.$$

Molybdenum–99 itself is produced in nuclear reactors.

> ### Chem Vocab
>
> Nuclear fission is a process in which a heavy nucleus (usually uranium or plutonium) splits into two parts, releasing neutrons and energy.

Nuclear Fission

In nuclear fission, big atoms break apart into smaller atoms. In nuclear fusion, the opposite occurs—small atoms combine together to make larger atoms. In both processes, however, a very large amount of energy is released. Thus, both processes can be used to make atomic bombs. Controlled fission has been used for over fifty years in the commercial production of electricity. There are serious technological difficulties that must be overcome, however, to have controlled fusion.

Uranium consists of about 99.3 percent U–238 and seven-tenths of 1 percent of U–235. It is U–235, however, that is fissionable. Fission is triggered by the absorption of a neutron to make U–236, which then breaks into two unsymmetrical fragments. There are various combinations of fission fragments possible. The following equations illustrate a few of the combinations:

➤ $^{235}_{92}\text{U} + ^{1}_{0}\text{n} \rightarrow ^{236}_{92}\text{U} \rightarrow ^{140}_{56}\text{Ba} + ^{94}_{36}\text{Kr} + 2^{1}_{0}\text{n}$

➤ $^{235}_{92}\text{U} + ^{1}_{0}\text{n} \rightarrow ^{236}_{92}\text{U} \rightarrow ^{95}_{38}\text{Sr} + ^{139}_{54}\text{Xe} + 2^{1}_{0}\text{n}$

➤ $^{235}_{92}\text{U} + ^{1}_{0}\text{n} \rightarrow ^{236}_{92}\text{U} \rightarrow ^{90}_{37}\text{Rb} + ^{143}_{55}\text{Cs} + 3^{1}_{0}\text{n}$

Notice that each fission event also results in the release of additional neutrons, an average of about 2.5 neutrons per fission.

The release of additional neutrons is the reason a chain reaction can take place. The neutrons are quickly absorbed by surrounding uranium atoms so that the rate of fission increases exponentially.

In a power plant the neutron-absorbing control rods are embedded in the uranium core to control the rate of fission. In an atom bomb there must be a critical mass of uranium–235 so that neutrons do not simply escape to the surroundings. Bombs, therefore, employ uranium that is highly enriched in U–235. Nuclear power plants also use an enriched form of uranium, but only enriched from 0.7 percent to a few percent U–235. If an explosion occurs in a nuclear power plant, it is an explosion of the steam boilers, not a nuclear explosion of the reactor core. There is always the potential for the accidental release of radiation from the power plant, but there is no potential for an atomic bomb explosion.

Chem Vocab

A nuclear chain reaction is a process in which neutrons that are released during the fission of one atom are subsequently absorbed by surrounding atoms, thereby causing them to undergo fission also. The number of atoms undergoing fission at each step increases exponentially.

Important Point

Contrary to public perception, nuclear power plants cannot blow up like atom bombs. The steam that is produced is under pressure and could blow up, but that is not an atomic explosion. Reactor cores can melt down, resulting in a release of radiation, but that is not an atomic explosion either.

Plutonium–239 is also fissionable. Plutonium is preferred for nuclear weapons for at least two reasons:

➤ It is easier to produce plutonium–239 from uranium–238 and then to separate plutonium from uranium than it is to enrich uranium.

➤ The critical mass of plutonium–239 necessary to fashion a nuclear warhead is less than the critical mass of uranium–235 that is necessary; therefore, smaller warheads can be made.

For these reasons plutonium was produced and stockpiled in ton quantities during the Cold War. Because of fears of it being diverted by terrorists or hostile nations into nuclear weapons, the supply of plutonium is tightly controlled by the military; plutonium is not used in civilian power plants.

There are pros and cons to using nuclear power. One of its advantages is that it can replace fossil fuels. Since no combustion of fuel is taking place, nuclear power does not contribute to global warming. Because fission produces a considerable amount of energy, a little uranium goes a long way, and the world's recoverable supply of uranium should last at least hundreds of years.

On the other hand, nuclear power has lost favor with the general public after highly publicized events at power plants: the Three Mile Island incident in 1979, the accident at Chernobyl in the Ukraine in 1986, and most recently the 2011 damage to Japan's Fukushima Daiichi nuclear power plant following the 9.0-magnitude earthquake and tsunami. One of the biggest controversies about nuclear power is that all of the products of fission are radioactive substances. Until a solution is found for recycling or storing the wastes for a very long period of time, the future of nuclear power in the United States remains uncertain. In addition, fear is ever present that radioactive or fissile materials could fall into the hands of terrorist groups or nations hostile to the United States.

The Aftermath of Fukushima

Before the meltdown at the Fukushima Daiichi power plant, Japan's energy policy was to expand nuclear power so that over 50 percent of its electricity would come from nuclear plants by 2030. Japan had planned to build fourteen more nuclear reactors, but that plan has changed. Instead, Japan is now looking at developing renewable energy resources like wind turbines, solar power, and biomass.

In the United States, a comprehensive review was ordered of the safety of all of the nuclear power plants operating in the United States. Of particular concern are the Diablo Canyon and San Onofre nuclear power plants that straddle the San Andreas Fault in California.

In other countries, China at least temporarily suspended all new approvals of nuclear power plants. Germany shut down its oldest nuclear reactors until the safety of continuing to use them could be determined. Throughout Europe nuclear reactors were inspected for safety.

Nuclear power is not going to disappear from the world scene, but it remains uncertain whether or not it will ever become a larger player on the energy field.

Nuclear Fusion

In nuclear fusion, two very small atoms, usually isotopes of hydrogen, combine to make a larger atom. A common example is the fusion of deuterium ($_1^2\text{H}$) and tritium ($_1^3\text{H}$) to make helium, as shown:

$$_1^2\text{H} + _1^3\text{H} \rightarrow _2^4\text{He} + _0^1\text{n}$$

The energy released in fusion per gram of hydrogen is roughly 10 times more than the amount of energy released in fission per gram of uranium.

A commercial nuclear fusion power plant would be an excellent source of energy. The supply of deuterium in the world's oceans would last many thousands of years. Tritium can be produced artificially. The helium that is produced is harmless; there are no radioactive waste products produced by fusion. Fusion is a very clean form of energy. Unfortunately, even after more than half a century of research, the technology necessary to build and operate a fusion power plant has eluded researchers. Whether or not fusion plants will ever be built is an unanswered question.

Chem Vocab

Nuclear fusion is a process in which light nuclei (usually isotopes of hydrogen) combine to form a heavier nucleus, resulting in the release energy.

Historical Note

Research into the development of nuclear fusion reactors began in the late 1950s with high hopes of reactors going online by the 1980s. The technological difficulties, however, have proven to be insurmountable.

Production of Superheavy Elements

Superheavy elements are defined as the transactinide elements—elements with an atomic number greater than 103. The current production of new superheavy elements takes place mostly at four locations around the world: the Lawrence Berkeley National Laboratory (LBNL) at Berkeley, California; the GSI Helmholtz Centre for Heavy Ion Research (GSI) at Darmstadt, Germany; the Joint Institute for Nuclear Research (JINR) at Dubna, Russia; and the Superheavy Element Laboratory of the Institute of Physical and Chemical Research (RIKEN) at Nishina, Japan. As of 2012, the discoveries of elements 113–118 have all been reported. The discoveries of four of these elements, however, are still awaiting independent confirmation from the other laboratories.

Chem Vocab

Superheavy elements are elements with an atomic number greater than 103. Currently, the discoveries of fifteen such elements have been reported.

As an example of new element production, just three atoms of element 118 were synthesized by a joint team of American and German scientists working together at Darmstadt. As shown in the following equation, a californium–249 target was bombarded by a beam of calcium–48 ions:

$$_{20}^{48}\text{Ca} + {}_{98}^{249}\text{H} \rightarrow {}^{294}118 + 3\,{}_{0}^{1}\text{n}.$$

Although element 118's position in the periodic table places it in the family of noble gases, the boiling point of element 118 is predicted to be above room temperature. If that proves to be true, then element 118 will be a noble liquid instead of a noble gas. As of early 2012, research was continuing into producing elements heavier than element 118. Presumably, element 119 will be an alkali metal, and element 120 will be an alkaline earth.

Conclusion

Nuclear processes provide humankind with a double-edged sword. On one hand, there are many useful applications of radioactive substances in science and medicine. Nuclear power is, and will continue to be, an important source of energy. On the other hand, there is always the danger of radioactive or fissile materials being used to threaten people's lives. No one can make radioactive or fissile materials just go away. Hopefully, wisdom will prevail, and peaceful applications of nuclear materials will dominate their use.

PART THREE

The Organization of Matter

Periodic Table of the Elements

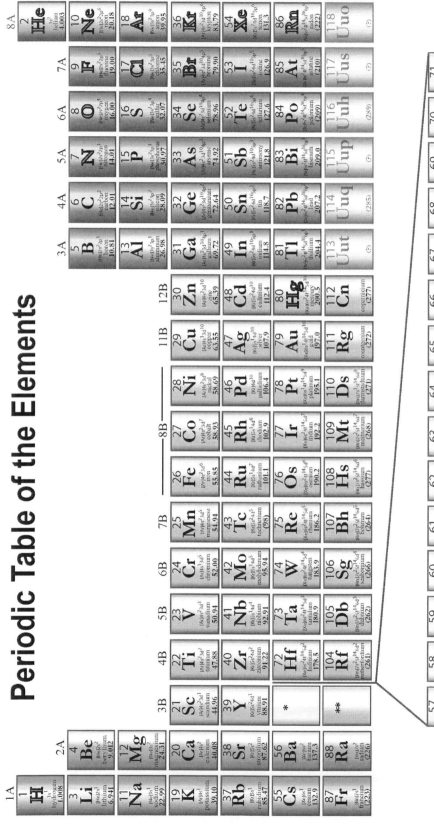

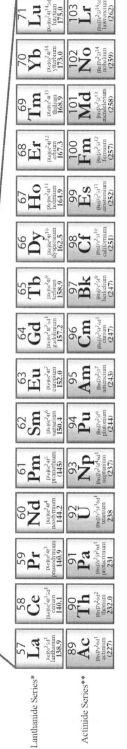

Lanthanide Series*

Actinide Series**

Los Alamos National Laboratory Chemistry Division

© Copyright 2011 Los Alamos National Security, LLC. All rights reserved. The public may copy and use this information without charge, provided that this Notice and any statement of authorship are reproduced on all copies. Neither the Government nor LANS makes any warranty, express or implied, or assumes any liability or responsibility for the use of this information.

Figure 15.1 – Periodic table

CHAPTER 15

The Periodic Table of the Elements

In This Chapter

➤ Discovery and naming of elements

➤ Transuranium elements

➤ Future of the periodic table

The chemical elements are frequently arranged into a table that has its origins in the work of the Russian chemist Dmitri Mendeleev (1834–1907), who published his version of the table in 1869. The periodic table is arranged into rows and columns, with the elements listed in order of increasing atomic number. The table is called *periodic* because there are regular trends in chemical and physical properties of the elements either descending a column or traversing a row from left to right.

Historical Note

In addition to his work in chemistry, Dmitri Mendeleev was also responsible for introducing the metric system to Russia.

Families of Elements

The elements fall into eighteen vertical columns called families or groups. The general organization of the periodic table is as follows:

> Most of the elements are metals and are located on the left side of the table.

> The elements on the right side are nonmetals.

> There is a group of elements called semimetals, or metalloids, found in-between the metals and nonmetals, marking the transition from metals to nonmetals. These elements are boron (B), silicon (Si), germanium (Ge), arsenic (As), antimony (Sb), tellurium (Te), polonium (Po), and astatine (At).

Important Point

The periodic table of the elements is the single most important unifying principle in chemistry.

Historical Note

Mendeleev predicted the existence of a few elements that had not been discovered yet because there were clearly holes in otherwise well-known families of elements. He did not predict the existence of the noble gases, however, because the entire family was unknown. There was no reason in 1869 to suspect that an entire column of the periodic table was in fact missing.

Some of the families of elements have names:

> The alkali metals—lithium, sodium, potassium, rubidium, cesium, and francium—are located in column 1 (hydrogen is in column 1, but it is not a metal)

> The alkaline earths—beryllium, magnesium, calcium, strontium, barium, and radium—are located in column 2

> The noble gases—helium, neon, argon, krypton, xenon, and radon—are located in the last column of the table

> The halogens—fluorine, chlorine, bromine, iodine, and astatine—occupy the column immediately preceding the noble gases

The elements copper (Cu), silver (Ag), and gold (Au) form a family that sometimes is called the coinage metals because of their widespread use in coins dating from ancient times. And the elements in the ten columns headed by scandium (Sc, number 21) to zinc (Zn, number 30), are called transition metals because as one moves across any of their rows from left to right it can be seen that the properties of the elements are making a transition from purely metallic to purely nonmetallic.

Important Point

Rows of elements in the periodic table are called periods. Columns of elements are called groups or families. As new elements are discovered, new rows of the periodic table will be filled. According to everything scientists know about elements, however, there will never be a new column added to the table.

Rows in the table are called periods. Just as there are trends in properties descending a column, there are trends in properties traversing a row from left to right.

The two rows at the bottom of the periodic table that have been separated from the rest of the table are called lanthanides and actinides. The lanthanides, the elements in the first row, from cerium (Ce) to lutetium (Lu), are called the lanthanides because their properties are similar to those of lanthanum (La). Historically, they have also been called the rare earths because they were believed to be much rarer than other metals. The elements in the second row, from thorium (Th) to lawrencium (Lr), are called the actinides because they are similar to actinium (Ac). Notice that the atomic numbers of the lanthanides all fit between numbers 57 (La) and 72 (Hf). The atomic numbers of the actinides all fit between numbers 89 (Ac) and 104 (Rf).

Important Point

With the exception of the noble gases, which occupy the last column of the periodic table and are exceptionally nonreactive, the most chemically reactive elements are located both at the far left side and the far right side of the table. The least reactive elements tend to be located in the central part of the table.

Properties of Metals, Nonmetals, and Metalloids

The importance of families of elements is that elements in the same family tend to have similar chemical and physical properties. Having similar chemical properties means that they form compounds with similar formulas. For example, the oxides of the alkali metals are Li_2O, Na_2O, K_2O, Rb_2O, Cs_2O, and Fr_2O, and the oxides of the alkaline earths are BeO, MgO, CaO, SrO, BaO, and RaO. Having similar physical properties means that there are regular trends in characteristics such as melting point, boiling point, and density.

Chem Vocab

A triad is a group of three elements with very similar chemical and physical properties. The properties of the element in the middle of a triad are often the average of the properties of the elements on either side of it. Triads usually are found in the same column. Sometimes, however, triads are found in the same row, as is the case with iron, cobalt, and nickel.

At the time Mendeleev developed the periodic table, a lot fewer elements were known. He often based his organization on what were called triads of elements—groups of three elements in the same vertical column with extremely similar properties. Examples of triads include vanadium, niobium, and tantalum in group VA; sulfur, selenium, and tellurium in group VIB; and chlorine, bromine, and iodine in group VIIB.

In the middle of the table, triads are grouped horizontally. Three columns are labeled group VIIIA because of the horizontal relationships. An example of a horizontal triad is iron, cobalt, and nickel. Sometimes there are also diagonal relationships. For example, beryllium is more similar to aluminum than it is to magnesium.

Mendeleev, of course, had no idea why such vertical, horizontal, or diagonal relationships existed. An explanation of those relationships had to wait for the discovery of electrons and atomic nuclei, and for the development of a detailed theory of the arrangements electrons assumed in atoms. Mendeleev lived long enough to witness the discoveries of the electron and proton, but died before the full story was unraveled.

We can draw a large number of conclusions about an element's chemical and physical properties from the position of that element in the periodic table. The following list gives some examples:

➤ Metals conduct heat and electricity, nonmetals are insulators, and semimetals (metalloids) are semiconductors of electricity.

➤ Metals are lustrous (shiny); nonmetals tend to be dull in appearance.

➤ Metals are malleable; they can be hammered and shaped into different forms. Nonmetals tend to be brittle in solid form.

➤ Metals are ductile; they can be drawn into wires.

Chem Vocab

Ductile means being able to be easily drawn into the form of a wire.

Malleable means having a shape that is easily changed. In the case of metals, malleable means they can be hammered into sheets.

➤ Metals form positive ions; nonmetals form negative ions. Noble gases are relatively nonreactive.

➤ There are trends in the melting point, boiling point, size of atoms, and other properties that vary with location in the periodic table.

➤ As a general rule, metalloids tend to have the physical characteristics of metals and the chemical properties of nonmetals. Metalloids can form both positive and negative ions.

We will often refer to the periodic table when discussing properties of various elements. Also, we will look at the periodic table in more detail later when we explain how the arrangement of electrons in an atom helps us understand the chemical and physical properties of an element.

Important Point

Gold is the most malleable of all metals. One sheet of gold leaf can be hundreds of times thinner than a human hair.

Important Point

The most important property of a metal is its ability to conduct electricity.

The Discovery and Naming of Elements

Several elements were never officially discovered, or at least their discoverers are unknown in the annals of history. These elements date to a time period between prehistoric times and the Middle Ages. Their names were originally in Latin, and their symbols today still reflect that origin. Lead, for example, was known as *plumbum*, hence its symbol Pb. Until fairly recently, most pipes were made of lead, from which was derived the word *plumbing*.

In ancient and medieval times, sodium and potassium were known only in compounds like rock salt (NaCl) and potash (K_2CO_3). Sodium and potassium were not actually isolated as elements until the early 1800s. Other elements, such as carbon and sulfur (S), were known to ancient people but were not recognized by them as being elements.

Historical Note

The following elements were known in ancient times:

➤ Antimony (Sb, stibium)

➤ Copper (Cu, cuprum)

➤ Gold (Au, aurum)

➤ Lead (Pb, plumbum)

➤ Mercury (Hg, hydrargyrum)

➤ Potassium (K, kalium)

➤ Silver (Ag, argentum)

➤ Sodium (Na, natrium)

➤ Tin (Sn, stannum)

➤ Iron (Fe, ferrum)

Beginning in the Middle Ages, alchemists began to discover new elements, including arsenic (As), bismuth (Bi), and phosphorus (P). The pace at which new elements were discovered accelerated in the eighteenth century, and many elements were added to the periodic table during the nineteenth and twentieth centuries. Until the twentieth century, all new elements had been discovered in naturally occurring materials.

In the 1930s, however, the first element to be made artificially, technetium, was identified by a scientific team working under the direction of the Italian-American physicist Emilio Segrè (1905–89). The name *technetium* was chosen to reflect the fact that it was produced by technical, or artificial, means. Also in the late 1930s, scientists at Berkeley, California, began discovering elements heavier than uranium, the transuranium elements. New elements are still being synthesized in laboratories located in the United States, Germany, Russia, and Japan.

In the 1940s, element 61 was discovered among the fission products produced in a nuclear reactor. Element 61 was named *promethium* after the Greek figure in mythology, Prometheus, who reputedly stole fire from the gods.

Historical Note

It is common to think of the first ninety-two elements in the periodic table as occurring in nature. That is almost true—chemists expected to find natural sources of elements 43 and 61 but finally realized that those two elements must be made artificially. It is true, however, that element 92 (uranium) is the heaviest element that occurs in nature if we ignore the few atoms of plutonium that sometimes occur in uranium deposits.

How Elements Are Named

The tradition in science is that the discoverer has the privilege of bestowing a name and symbol upon his (or her) discovery. Some elements have been named according to their appearance or properties; some have been named for mythological gods or goddesses, or for planets or asteroids. The names of several elements have been derived from geography—cities, states, countries, regions, or continents. In modern times it has become common to name elements after famous scientists who may, or may not, have been an element's discoverer.

Table 15.2 –Origins of Names of Sample Elements

Element	Year of Discovery	Discoverer	Origin of Name
Cobalt (#27, Co)	1735	G. Brandt, Sweden	German for "goblin"
Uranium (#92, U)	1789	M. H. Klaproth, Germany	Named after the planet Uranus
Vanadium (#23, V)	1801 rediscovered in 1831	N. G. Selfström, Sweden (1831)	Named after the Scandinavian goddess Vanadis
Iridium (#77, Ir)	1803	S. Tennant, England	Latin for "rainbow"
Holmium (#67, Ho)	1878	P. T. Cleve, Sweden	Named after Stockholm
Gadolinium (#64, Gd)	1880	J. C. Galissard de Marignac, Switzerland	Named after the Finnish chemist, J. Gadolin
Europium (#63, Eu)	1901	E. A. Demarcay, France	Named after Europe
Curium (#96, Cm)	1944	Glenn Seaborg *et al*, California	Named after Pierre and Marie Curie
Berkelium (#97, Bk)	1949	Glenn Seaborg *et al*, California	Named after the city and university

Names and spellings of elements may vary with language, but symbols are universal. Tungsten, for example, is *wolfram* in Europe, but *W* is the universal symbol. Originally, elements could have either one or two letters in their symbols. The practice today, however, is to always use two letters. A symbol is either a single capital letter, or else two letters—a capital letter followed by a lower case letter. Inspection of the periodic table will show unnamed elements 113–118 with symbols like *Unt, Unq*, etc. These symbols are only temporary and represent Latin abbreviations for the numbers 113, 114, 115, and so forth, and will be replaced when these elements eventually are given permanent names.

Relative Abundance of Elements

The most abundant element in the universe is the lightest element—hydrogen—followed by the second lightest element—helium. The Sun, for example, is about three-fourths hydrogen, one-fourth helium, and has only traces of all the other elements.

On Earth, the most abundant elements are oxygen, which is almost 21 percent of the atmosphere by volume and almost 46 percent of Earth's crust by mass; and silicon, which is about 27 percent of Earth's crust. Together, oxygen and silicon comprise about 73 percent of Earth's crust. This should not be too surprising, considering that quartz and sand (which is just decomposed quartz) are made of oxygen and silicon, and both substances are widespread throughout the world.

The next most abundant elements on Earth are aluminum, iron, calcium, magnesium, sodium, and potassium, which together make up about 26 percent of Earth's crust, with a large amount of sodium and potassium also found in seawater. Their abundance in part explains why their names should sound so familiar. Of course, other reasons for their familiar names include the widespread use of aluminum and iron in today's economy, plus the fact that iron, calcium, magnesium, sodium, and potassium are some of the elements that are essential to human health.

Important Point

The most abundant elements in Earth's crust are oxygen, which constitutes almost one-half of the crust, and silicon, which constitutes more than one-quarter. The least abundant naturally occurring elements in Earth's crust are ruthenium and rhodium—both exist at concentrations of about 0.1 ppb.

The elements listed so far constitute about 99 percent of Earth's crust. That means all the remaining naturally occurring elements together constitute the remaining 1 percent! Compared to the eight most abundant elements, most of the remaining elements are found in only a small amount on Earth.

Fortunately, many elements that are useful in industry are sufficiently concentrated in ores so that mining them is economical. Chromium, nickel, zinc, copper, and lead are much less abundant than aluminum or iron, but they are still very economical to mine. Their low abundance, however, is good cause for recycling them as much as possible.

Some important elements are found in only trace quantities, but the demand for them is great. The law of supply and demand is what makes them so expensive. Included in this group are elements such as molybdenum, tungsten, silver, mercury, palladium, platinum, and gold, all of which are elements that definitely should be recycled. It is probably true that virtually 100 percent of the gold that has ever been recovered throughout all of human history has been recycled.

Of course, the most essential element for life as we know it is carbon. Carbon makes up about 0.034 percent of the atmosphere by volume and only 0.018 percent of Earth's crust by weight. There are over a dozen elements that are more common on Earth than carbon, but without carbon life could not exist.

Important Point

Carbon may not be the most abundant element, but its chemistry is certainly the most versatile of all elements. No other element can form the complex variety of chemical compounds required by living organisms.

The Not-So-Rare Earths

It surprises most people to learn that several of the so-called rare earth elements are not actually that rare compared to much more familiar elements. Neodymium, praseodymium, samarium, gadolinium, dysprosium, erbium, and ytterbium are all more abundant than more familiar elements like bromine, uranium, or tin. Europium, holmium, terbium, lutetium, and thulium are more abundant than iodine, silver, or mercury. Yet few people have even heard of most of the rare earths. The reason is that rare earths tend not to concentrate in large ore deposits in the way that better known metals do. Historically there have been fewer profits to be made from mining rare earth elements, and there have been fewer applications developed for them in industry.

That situation has changed in recent years, however. The discovery of major deposits of rare earths in China has led to extensive research into their properties and the development of applications in technologies such as magnets and hybrid vehicles.

Historical Note

Between 1965 and 1995 the Mountain Pass Mine in southern California supplied most of the rare earth metals used around the world. Competition from China, however, led to the closure of the mine. In recent years the worldwide demand for rare earths has become so great that the Mountain Pass Mine reopened in 2011.

Relative Abundance of Elements in the Human Body

Because biological processes tend to concentrate certain elements in living organisms, the relative abundance of elements in the human body is much different from what they are in the atmosphere, in the oceans, or in Earth's crust. By weight, 63 percent of the body is hydrogen, 24 percent is oxygen, and about 10 percent is carbon. That leaves the rest of the elements found in the body to constitute only about 3 percent of the body. Nitrogen is the next most important, followed mainly by phosphorus, iron, copper, calcium, magnesium, sodium, potassium, iodine, and zinc.

Table 15.3 – Elements Comprising the Human Body

Element	Percentage by Mass
Oxygen (O)	61
Carbon (C)	23%
Hydrogen (H)	10%
Nitrogen (N)	2.6%
Calcium (Ca)	1.4%
Phosphorus (P)	1.1%
Sulfur (S)	0.20%
Potassium (K)	0.20%
Sodium (Na)	0.14%
Chlorine (Cl)	0.12%
Magnesium (Mg)	0.027%
All other elements	< 0.3%

The Transuranium Elements

A copy of the periodic table produced in 1930 would have shown ninety elements. All of the elements from hydrogen through uranium were known with the exception of elements 43 and 61. Elements 43 (technetium) and 61 (promethium) were produced artificially, but that still did not mean that the periodic table ended with uranium. By the time promethium had been discovered, the first elements lying beyond uranium—the transuranium elements— were beginning to be discovered.

A copy of the periodic table produced in 2012 shows a total of 118 elements. The discoveries of elements 93 through 112, 114 and 116 have been sufficiently confirmed that they have been allowed to give names to their discoveries. (The International Union for Pure and Applied Chemistry [IUPAC] has the final say in naming new elements.)

Important Point

There are twenty-eight elements that do not occur in nature and have been produced artificially. That constitutes almost one-quarter of the periodic table!

Although the discoveries of elements 113, 115, 117, and 118 have been reported, their discoveries are waiting independent confirmation from other laboratories. When experiments confirm that these elements really have been made, then their discoverers will be allowed to recommend names and symbols for them.

The following list gives the atomic numbers of transuranium elements 93-112, 114, and 116, their names and symbols, and the people or places after whom they have been named:

Atomic number	Name of element	Symbol	Person or place after whom named
93	neptunium	Np	the planet Neptune
94	plutonium	Pu	the planet Pluto
95	americium	Am	America
96	curium	Cm	Pierre (French) and Marie Curie (Polish)
97	berkelium	Bk	University of California, Berkeley
98	californium	Cf	the university and the state
99	einsteinium	Es	Albert Einstein (German)
100	fermium	Fm	Enrico Fermi (Italian)
101	mendelevium	Md	Dmitri Mendeleeev (Russian)
102	nobelium	No	Alfred Nobel (Swedish)
103	lawrencium	Lr	Ernest Lawrence (American)
104	rutherfordium	Rf	Ernest Rutherford (British)
105	dubnium	Db	Russian city of Dubna
106	seaborgium	Sg	Glenn Seaborg (American)
107	bohrium	Bh	Niels Bohr (Danish)
108	hassium	Hs	German state of Hesse
109	meitnerium	Mt	Lise Meitner (Austrian)
110	darmstadtium	Ds	German city of Darmstadt
111	roentgenium	Rg	Wilhelm Röentgen (German)
112	copernicium	Cp	Nicholas Copernicus (Polish)
114	flerovium	FL	Flerov Laboratory, Dubna
116	livermorium	Lv	Lawrence Livermore National Laboratory

Beginning with einsteinium, none of these elements has ever been made in more than very tiny quantities. In some cases only a few atoms have ever been made. Also, all of their

---

Historical Note

Lise Meitner (1878–1968) was an Austrian-born physicist who spent most of her scientific career working with the German chemist Otto Hahn (1879–1968). Together in the 1930s, they discovered nuclear fission. Meitner was Jewish, however, and in the prevailing atmosphere of Nazi Germany, Hahn could not give Meitner credit for her contribution. When Hahn was awarded the 1944 Nobel Prize in Chemistry, the prize was not shared with Meitner. Lise Meitner, however, was later honored by a place in the periodic table—element 109.

---

isotopes are radioactive, and many of their isotopes have very short half-lives—in many cases less than a second. Therefore, no one has ever actually seen these elements. The best we can do is to say that their presence was detected with laboratory instruments before they decayed.

Conclusion: The Future of the Periodic Table

Is the periodic table complete? Are there any more new elements to be discovered? Scientists believe there are. As of early 2012, a team at Darmstadt, Germany, is trying to synthesize element 120.

Theoretically, there is no known reason why the periodic table cannot be extended indefinitely. Theory even suggests that there may be elements much heavier than the ones that have already been produced that will have significantly longer half-lives. Of course, when considering heavy elements, a half-life of one second is considered long. It may be, however, that there are heavy elements waiting to be discovered that will have half-lives measured in years, even thousands of years.

 # Early Models of the Atom

In This Chapter

- ➤ Democritus
- ➤ John Dalton's model
- ➤ J. J. Thomson's model
- ➤ Ernest Rutherford's model

In the previous chapter we said that all matter is made of atoms. That means all the matter that is in stars, all the matter that lies in the vast expanses of space between stars, all the matter that makes up planets and moons, all the matter that is in the atmospheres, rocks, and oceans of those planets and moons, and all the matter that makes up all living organisms, including humans. There is no difference between the hydrogen atoms that make up three-fourths of our Sun, the hydrogen atoms in the waters of Earth's deepest oceans, and the hydrogen atoms that bind together the strands of DNA that are inside our body's cells.

But what is an atom? It may surprise you—in fact, it may astonish you—to find out that nobody really knows what an atom is. No one has ever seen an atom. Atoms are just too tiny to be seen, even with the aid of the best microscopes in the world. No one even knows how to accurately describe an atom. The best anyone can do is to make a model of an atom and then to explain the atom's properties in terms of the model.

Important Point

A 1-liter bottle of water contains 100,000,000,000,000,000,000,000,000 (1×10^{26}) atoms, so obviously any one atom is really infinitesimally small!.

This chapter is about the different models of the atom that have been proposed over the centuries. Some models are relatively simple, some are relatively complex, but none are completely correct. Depending upon the properties of atoms in which one is interested, however, picturing atoms according to one of the accepted models can be very useful.

Democritus and the Ancient Greeks

Ancient Greek philosophers were interested in trying to understand nature. They wondered if matter was infinitely divisible or if there is some smallest piece that cannot be divided any further? *Infinitely divisible* means that you could tear a sheet of paper in half, tear one-half in half again, tear a piece in half again, and so on forever, at least in theory. There would never be a piece that could not be divided. (Obviously, it would be impossible to actually try to do this, but the Greeks were not into doing experiments anyway, so doing this as a "thought" experiment sufficed.)

Democritus (460?–370? BC) was a Greek philosopher who was a proponent of the idea that matter is composed of indivisible particles called atoms (from the Greek word *atomos*, which means "not able to be cut"). Democritus's ideas, however, did not prevail. Aristotle (384–322 BC) was extremely influential in matters of natural philosophy. He rejected Democritus's theory of atoms, and the idea of matter being made of atoms was set aside for more than 1,500 years. Although great scientists like Galileo and Newton believed in atoms, they really could not offer any support for the idea, so atomic theory continued to lie dormant.

Historical Note

Because John Dalton revived the ancient notion of the existence of atoms, he is often referred to as the father of modern atomic theory.

John Dalton, the Father of Modern Atomic Theory

John Dalton (1766–1844) was an English chemist who was well respected by his peers. In 1803, Dalton revived Democritus's theory about atoms. Dalton believed that assuming matter is made of atoms offered the best explanation for the laws of chemistry that had recently been discovered, namely the law of conservation of mass in chemical reactions, the law of definite proportions of elements in compounds, and the law of multiple proportions in the ways elements can combine.

Conservation of Mass

We discussed the law of conservation of mass in Chapter 7. If you recall, this law states that mass is neither created nor destroyed during chemical reactions. For example, consider the combustion of butane (C_4H_{10}) in air, as shown:

$$2\ C_4H_{10}\ (g) + 13\ O_2\ (g) \rightarrow 8\ CO_2\ (g) + 10\ H_2O\ (g).$$

Suppose we begin with 116.28 grams of butane and 416.00 grams of oxygen gas and let them react until no butane or oxygen molecules remain. If we collect the carbon dioxide and water that form, we will find that we have 352.08 grams of CO_2 and 180.20 grams of H_2O. We started with 532.28 grams of reactants (116.28 + 416.00), and we get 532.28 grams of products (352.08 + 180.20).

Dalton explained the law of conservation of mass basically by saying that it is really conservation of atoms that is occurring. Atoms are neither created nor destroyed in ordinary chemical reactions (ones that do not involve changes in the nuclei); they are just rearranged to form compounds that are different from the starting materials. Since the masses of individual atoms do not change in the process, mass is conserved.

Of course, it should be pointed out that we have done this calculation to only two decimal places. Realistically, that is probably the best precision that people in Dalton's time could have obtained. If we could, however, weigh the reactants and products to eight or more decimal places, we would find that the total mass in this reaction actually decreases slightly. But even today, most laboratory balances weigh materials to a precision of only one to four decimal places, so changes in weight in the eighth, ninth, or tenth decimal place still would go unnoticed. This small change in mass is important in nuclear reactions but is generally ignored in ordinary chemical reactions.

Chem Vocab

To conserve something means that its properties do not change. Chemists most often speak of conservation of mass, energy, and electrical charge.

Definite Proportions

The law of definite proportions says that the elements in compounds combine in fixed ratios. For example, in water the ratio by mass of oxygen to hydrogen is about 8 to 1. In carbon dioxide the ratio of oxygen to carbon is about 8 to 3. We would never find a water molecule that had a ratio of oxygen to hydrogen of, say, 7 to 1, or 9 to 1, or any ratio other than 8 to 1. Similarly, we would never find a molecule of carbon dioxide with a ratio of oxygen to carbon of anything other than 8 to 3.

Dalton explained the law of definite proportions by saying that only whole numbers of atoms can combine to form compounds. In water there are always two hydrogen atoms for every oxygen atom. In carbon dioxide there are always two oxygen atoms for every carbon atom. Since each kind of atom has its own characteristic mass, the ratios of the masses of the elements in any given compound will always be the same.

Important Point

By the law of definite proportions, every water molecule in the universe has the formula H_2O. Without exception, every water molecule has two atoms of hydrogen and one atom of oxygen. Any other combination of hydrogen and oxygen and it is not a water molecule.

Multiple Proportions

The law of multiple proportions says that two or more elements sometimes can combine to form different compounds. Each compound alone still satisfies the law of definite proportions; the different combining ratios refer to different compounds. For example, carbon and oxygen can combine to form carbon monoxide (CO) or carbon dioxide (CO_2). In CO the mass ratio of oxygen to carbon is 4 to 3; in CO_2 the ratio is 8 to 3. Different proportions are possible, but each one means a different compound with different chemical and physical properties. Many such examples exist: $FeCl_2$ and $FeCl_3$; N_2O, N_2O_4 and N_2O_5; CH_4, C_2H_6, C_5H_{10}, C_6H_6; and many (millions, in fact!) more.

Chem Vocab

The term *definite proportions* means that the elements in a given compound are always combined in the same proportion by mass.

The term *multiple proportions* means that any time two elements can form more than one compound, the different masses of one element that combine with the same mass of the other element are in the ratio of small whole numbers.

Dalton's Model of the Atom

Dalton had no way to study the properties or structure of individual atoms. His model is often referred to as the little hard sphere model of the atom. In Dalton's mind, atoms had no internal structure but were essentially like tiny ball bearings or marbles. What distinguished atoms of one element from atoms of a different element was relative mass. Consequently, from that time on it was common to list the known elements in order of relative mass.

Lists always began with hydrogen, the lightest element, which was assigned a mass of 1. From there, lithium would be next (since helium was not discovered until much later) with a mass of 7, beryllium with a mass of 9, boron with a mass of 11, carbon with a mass of 12, and so forth through uranium, the heaviest naturally occurring element with a mass of 238.

Dalton's little hard sphere model of the atom may seem primitive by today's standards, but it was an essential step in the evolution of chemical knowledge. Dalton's model persisted for almost one hundred years before anyone could think of any way to improve upon it. What is especially remarkable is that Dalton's theory was not completely accepted by the scientific community. Until 1900, there remained prominent physicists and chemists who continued to deny the existence of atoms. Actually, probably the most unsatisfying thing about Dalton's model is that it offered no explanation for the differences in chemical and physical properties that were observed among the elements. Even Dmitri Mendeleev, who, in 1869, developed the modern periodic table of the elements, could offer no explanation for the regular, or periodic, trends in the elements that were displayed in his periodic table. For that explanation, we must turn the clock forward to the events of the 1890s.

Historical Note

John Dalton reintroduced the notion of atoms in the first decade of the nineteenth century. For the next one hundred years, scientists hotly debated whether or not atoms even existed. It was not until the first decade of the twentieth century that atomic theory finally was fully accepted.

Table 16.1: The Classical Picture of Matter and Light
Before 1900, it was believed that matter has only a particle nature and light has only a wave nature.

	Particles	Waves
Matter	X	
Light		X

Classical Physics in the 1800s

During the 1800s, physicists believed that they had pretty well figured out the nature of matter and light. Matter consisted of particles, which are mostly characterized by their masses. Light consisted of waves, which are mostly characterized by their frequencies and energies. To nineteenth-century physicists, matter and light seemed to have little in common with each other (table 16.1). That picture would change in the last year of the century.

Henri Becquerel and the Discovery of Radioactivity

Antoine Henri Becquerel (1852–1908) came from a family of French physicists. X-rays had been discovered in December 1895 by the German physicist Wilhelm Roentgen (1845–1923). In 1896, Becquerel began his own series of experiments designed to find out what kinds of substances would emit X-rays. Becquerel discovered that uranium salts also emitted rays, but rays that differed from X-rays. (X-rays are a very high-energy form of light, whereas the rays Becquerel discovered proved to be charged particles.) Becquerel then collaborated with the husband-and-wife team of Pierre (1859–1906) and Marie (1867–1934) Curie in further investigating the nature and origin of these rays.

Chem Vocab

Ultraviolet light has a higher frequency (and a shorter wavelength) than visible light does. Higher frequency also means higher energy.

X-rays have a higher frequency and energy than ultraviolet light does.

Gamma rays have an even higher energy than X-rays do. Gamma rays are the most dangerous form of radiation.

Becquerel and the Curies concluded that the rays emanated from within the uranium itself. This result suggested that uranium atoms are not little, indestructible, hard spheres (as Dalton had suggested), but that they can in some way emit particles. In particular, they identified one kind of particle as having a negative charge. The notion that atoms have structure to them was born. It was Marie, in fact, who coined the term *radioactivity* to describe this phenomenon. For their work, Becquerel and the Curies shared the 1903 Nobel Prize in physics.

Historical Note

Pierre and Marie Curie shared the 1903 Nobel Prize in physics for their work explaining radioactivity that occurs naturally. Their daughter, Irene Joliot-Curie (1897–1956), and Irene's husband, Jean-Frédéric Joliot-Curie (1900–1958), shared the 1935 Nobel Prize in chemistry for their discovery of radioactivity that can be induced artificially.

J. J. Thomson and the Discovery of the Electron

Joseph John "J.J." Thomson (1856–1940) was an English physicist. In 1897, Thomson investigated the nature of "rays" that traveled between charged plates in a vacuum tube. What Thomson found was that the "rays" were not a form of light like X-rays, but that they had mass and a negative charge. Moving charged particles are deflected by both electric and magnetic fields. By varying the strength of the fields and measuring the deflections, Thomson was able to determine the charge–to-mass ratio of the particles. He found that the ratio was about 1,836 times greater than the charge to mass ratio of a hydrogen ion, which is positively charged.

Assuming that the charges had the same magnitude but were just opposite in sign, this result suggested that the masses of these particles were at least 1,830 times less than the mass of a hydrogen ion. Thomson concluded that the particles were constituents of atoms, striking the final blow to Dalton's idea of atoms being little hard spheres. Physicists had already coined the name *electron* to designate the particle presumed to carry the charge in electrical circuits, and Thomson adopted that name. Scientists then realized that the negatively charged products of radioactive decay that had been observed by Becquerel and the Curies were, in fact, electrons.

Thomson's Model of the Atom

No one actually discovered protons, but physicists knew that atoms were electrically neutral. If there were negatively charged particles in atoms, then there also had to be positively charged particles. These particles were named protons and were recognized to be responsible for the mass of atoms, the electrons contributing almost negligibly to an atom's mass.

Thomson proposed what he called a plum pudding model of the atom. English plum puddings contain plums, which are relatively large, and raisins, which are comparatively very small. Thomson said that the protons were like the plums and electrons were like the raisins. Just like the plums and raisins in a plum pudding are uniformly mixed together, Thomson suggested that protons and electrons are uniformly mixed together in atoms. Any atom would have to have an equal number of protons and electrons. Thus, a carbon atom with an atomic weight of 12 would have 12 protons and 12 electrons.

This model was not correct, of course. First of all, almost all atoms also contain neutrons (the only exception being the lightest isotope of hydrogen, H–1), but the neutron was not discovered until 1932, so it did not play a role in models of the atom during the first three decades of the twentieth century. Secondly, the protons and electrons in atoms are not uniformly distributed, but another experiment had to be done to figure that out. Thomson's discovery of the electron, however, earned him the 1906 Nobel Prize in physics.

Historical Note

J. J. Thomson was awarded the 1906 Nobel Prize in physics for his discovery of the electron, using an experimental method that demonstrated the particle nature of the electron. His son, George Paget Thomson, shared the 1937 Nobel Prize in physics for work that demonstrated the wave nature of the electron.

Ernest Rutherford and the Discovery of the Nucleus

Ernest Rutherford (1871–1937) was born in New Zealand, emigrated to Cambridge, England, where he was a student of J. J. Thomson, then moved to Montreal, Quebec, and finally returned to Cambridge. Sometimes called the founder of nuclear physics, Rutherford led a distinguished and productive career. He established that in radioactive decay there are at least two kinds of particles emitted—positively charged particles, which he called alpha particles, and the negatively charged particles discovered by Becquerel and the Curies, which he called beta particles. (Alpha particles were shown to be helium ions, while beta particles were shown to be electrons.)

Rutherford showed that uranium and thorium atoms undergo a series of successive decays and coined the term *half-life* to describe the time required for half a sample of a radioactive substance to decay. For this work, Rutherford was awarded the 1908 Nobel Prize in chemistry.

The Gold Foil Experiment

In 1911, Rutherford performed an experiment designed to test the validity of Thomson's plum pudding model of the atom. Rutherford bombarded a very thin gold foil with alpha particles. (Gold was chosen because gold is the most malleable of all metals, therefore it can be hammered into the thinnest possible sheets.) If the atoms in the gold foil really consisted of uniformly distributed protons and electrons, it would be reasonable to expect the positively charged alpha particles to be repelled by the positively charged protons in the gold atoms. In other words, the alpha particles should just bounce back from the foil.

What Rutherford found, however, was that most of the alpha particles passed right through the gold foil as though the foil were not even there. Only a few particles were deflected, some straight back out but mostly over a range of angles to the direction of the incident beam. It looked like the foil consisted mostly of empty space, not of a uniform distribution of matter. By carefully measuring the distribution of the angles of deflection of the alpha particles, Rutherford was able to conclude that the atoms were, in fact, mostly empty space but had a very, very tiny center of mass and positive charge.

The Nuclear Model of the Atom

Rutherford called the tiny center of mass and positive charge the atom's nucleus, a term borrowed from biologists who had been calling the centers of the cells of living organisms nuclei for almost a century. Rutherford replaced Thomson's plum pudding model of the atom with a nuclear model. In the nuclear model all of the protons and roughly half of the electrons were concentrated in the nucleus, and the remaining electrons were distributed in the (mostly empty) space outside the nucleus.

For example, it was known that carbon has an atomic weight of 12 and that six electrons could be stripped away to form an ion with a charge of +6. To account for this information, the nuclear model would suggest that a nucleus of carbon contains 12 protons (since protons would be responsible for the mass) and six electrons, giving the nucleus a net charge of

+6 (12 positive charges minus 6 negative charges). Because the atom has to be electrically neutral, there had to be an additional six electrons located in the region of space outside the nucleus.

Furthermore, Rutherford was able to measure the relative size of the nucleus and the space occupied by the six outer electrons (which would determine the size of the atom itself). He concluded that the diameter of the atom was about 10,000 times larger than the diameter of the nucleus. Since the diameter of a gold atom is on the order of 10^{-10} m, that would make the diameter of the nucleus on the order of about 10^{-14} m.

In ensuing years, scientists like Rutherford began to suspect that atoms contained a third, neutral particle, but the discovery of the neutron was two decades away. Once the neutron was discovered, the nuclear model became one in which the nucleus contained protons and neutrons, but no electrons. The number of protons corresponded to the charge on the nucleus, and the sum of the numbers of protons and neutrons determined the mass of the atom. The electrons were located only in the space outside the nucleus, the number of electrons in a neutral atom being equal to the number of protons. Eventually, the concept of the atomic number of an element was developed to correspond to the number of protons in an atom. The concept of the mass number of a specific isotope was developed to correspond to the sum of the numbers of protons and neutrons.

Conclusion

For many purposes in chemistry, the nuclear model of the atom is still a perfectly sufficient model for picturing atoms. It explains atomic numbers and atomic weights, allows for different isotopes of elements as well as different ions, and is all that is necessary to study mass relationships in chemical reactions. Physicists and chemists, however, were not satisfied just to say that the electrons are somewhere in the region of space outside the nucleus. A more detailed understanding of how the electrons are distributed was both the next subject of their investigations and the subject of our next chapter.

CHAPTER 17

Modern Models of the Atom

The work being described in this chapter mostly took place from about 1913 to 1926. The development of our modern model of the atom—the quantum mechanical model—is a remarkable achievement in the history of science. Much of twenty-first century science and technology depends on the properties of atoms and molecules as described in the quantum mechanical model of the atom.

Niels Bohr and the Planetary Model of the Atom

Niels Bohr (1885–1962) was a young Danish physicist who had just earned his PhD when in 1912 he went to England to work with Ernest Rutherford. Rutherford's nuclear model of the atom was still new and was the subject of controversy. The main difficulty with Rutherford's model was the electrostatic attraction between the positively charged protons in an atom's nucleus and the negatively charged electrons residing somewhere outside the nucleus. Why did the atom not collapse? The electrons should be quickly drawn into the nucleus. Under Rutherford's model, atoms should not even exist. The young physicist from Denmark decided to tackle the problem of explaining the stability of atoms–a formidable task indeed.

Chem Vocab

Electrostatic attractions are forces of attraction between positively and negatively charged particles.

Important Point

In the planetary model of the atom, an atom looks like a miniature solar system. Electrons orbit an atom's nucleus just like planets orbit the Sun.

In 1913, Bohr published what he called the planetary model of the atom because it resembled the solar system. Bohr envisioned the nucleus of an atom as being analogous to the Sun, and the electrons as being analogous to the planets revolving around the Sun. In effect, Bohr said that an atom is a miniature solar system.

Besides the huge difference in size between the solar system and an atom, there was another crucial difference between Bohr's planetary model of an atom and the planetary model of the solar system. In the solar system there is no restriction on how far away a planet can orbit the Sun. The average distance between Earth and the Sun, for example, is 93 million miles. There is no requirement, however, that Earth must be that distance. Earth could be 80 million miles from the Sun or 100 million miles, and it would continue to orbit the Sun just as it does now. Of course, whether or not liquid water can exist on Earth—and, therefore, life as we know it—does depend on Earth's distance from the Sun. Too close to the Sun, and water would exist only as a vapor. Too far away, and water would be frozen as ice. But that is another consideration. As far as the planet itself is concerned, any distance from the Sun would be satisfactory.

Quantum Theory

Bohr said that electrons orbiting an atom's nucleus are different in that an electron is constrained to have a specific energy and, therefore, to be a specific distance from the nucleus. Bohr was drawing on the idea of quantization of energy that had been formulated by the German physicist Max Planck (1858–1947) in 1900 and supported by Albert Einstein in 1905. Planck had said that in addition to its wavelike nature, light exhibits a particle-like nature and travels in little "packets" that have an energy given by the following equation:

$$E = h\nu = hc/\lambda,$$

where h is called Planck's constant and has a value of 6.63×10^{-34} J·s, the Greek letter ν (nu) is the frequency of light in units of cycles per second (or hertz), the Greek letter λ (lambda) is the wavelength of light in meters, and c is the speed of light in a vacuum (3.00×10^8 m/s).

Important Point

Planck's constant is a very tiny number. Because it is so tiny, we are not aware of quantization of energy in our everyday, macroscopic world. If Planck's constant were significantly larger, we would live in a quantum world, and life would be much different from what we are used to.

According to quantum theory, light is said to exhibit a dual nature—it has the properties of both particles and waves (figure 17.1). Light's wave nature has an easy analogy: we think of light waves as being similar to water waves. A fundamental difference, though, between light and any other kind of wave is that all other kinds of waves (water, sound, seismic waves, or waves on the strings of musical instruments) require a material medium through which to travel. Light is the only kind of wave that can travel through a vacuum. Because light is not a material substance, however, it is more difficult to picture its particle nature. We refer to light as traveling in packets of energy, or as corpuscles (like blood cells), or as photons, a term that was coined just to try to describe light's particle nature. None of these terms, however, really tells us what a particle of light really is.

Chem Vocab

A photon is the quantum, or unit, of light. A photon may be thought of as a stable, discrete particle that possesses no mass or electrical charge.

Figure 17.1: The Dual Nature of Light
After 1900, scientists believed that light has a dual nature—it possesses the properties of both particles and waves.

	Particles	Waves
Matter	X	
Light	X	X

Important Point

The explosion of the Death Star in *Star Wars VI: Return of the Jedi* may have been visually very dramatic, but the Ewoks who watched the explosion from their home planet would have heard nothing. Light propagates very well through the vacuum of space, but sound does not. Sound can only propagate through material media like air or water.

The energy of a body in orbit around a central object is given by the sum of the body's kinetic and potential energies and is determined by the distance the orbiting body is from the central object. Planets can have any energy, so they can be any distance from the Sun. Quantization of energy, however, means that electrons can have only certain energies, and, therefore, they can be only specific distances from the nucleus of an atom. (In theory, quantization also applies to objects as large as planets, but the effects are too miniscule to matter. For all practical purposes quantization only manifests itself in the microscopic world of atoms and molecules.) The development of quantum theory won Planck the 1918 Nobel Prize in physics.

The Planetary Model of the Atom

Applying his theory to the simplest atom, hydrogen, with its single electron, Bohr successfully derived an equation that gave the possible energies the electron could have. The orbit with the lowest energy—called the ground state of the atom—was the orbit closest to the nucleus. Orbits that are farther from the nucleus—called the excited states of the atom—were higher in energy. The electron could travel around the nucleus in one of several possible discrete orbits, but the electron could not travel around the nucleus in the space between these orbits.

Important Point

Generally speaking, electrons populate excited states of atoms for only very, very tiny fractions of a second. Most of the time, electrons will populate the lowest-energy states that are available to them.

In the case of an orbit being so far from the nucleus that there is no longer any real attraction between the electron and the nucleus, the electron could leave the atom altogether. In that case we would say that the atom has been ionized.

The electron could make virtually instantaneous transitions between the allowed orbits. Most of the time the electron would be expected to occupy the ground state of the atom. If the atom absorbed energy in the form of light or heat, however, the electron could possibly move to an orbit with a higher energy. To do so the energy of the light or heat absorbed would have to correspond exactly to the difference in energy between the two orbits. The electron could also fall back down from a higher energy orbit to a lower energy orbit. In doing so, light with energy that is exactly equal to the difference in energy between the two orbits would have to be emitted.

Bohr could calculate the frequencies, or wavelengths, of the light absorbed or emitted by hydrogen atoms and found that his calculated values agreed quite closely with experiment. It was a remarkable achievement and earned Bohr the 1922 Nobel Prize in physics. It is very common to picture atoms as having electrons orbiting the nuclei like little planets orbiting the Sun. In fact, the symbol for the Nuclear Regulatory Commission uses this picture.

As it turned out, however, Bohr's planetary model was not the final answer to the question of the electronic structure of atoms. For one thing, this equation did not work for neutral atoms of any of the other elements in the periodic table. When corrected for the higher charge on other nuclei, his equation did work for ions of elements that had only one electron, but not if they had two or more electrons. In addition, more careful experiments indicated that there were small discrepancies between the actual wavelengths of light observed in hydrogen's spectrum and the values calculated from the Bohr model. Even more damaging, the wavelengths were observed to change if the sample of hydrogen gas was placed in an electric or magnetic field, and the Bohr model could not explain those changes at all. It was back to the drawing board, but the beginning of World War I in 1914 delayed further investigation. Our story of the atom continues after the end of the war.

Historical Note

Niels Bohr was awarded the 1922 Nobel Prize in physics for his explanation of the structure of the atom. His son, Aage Niels Bohr, shared the 1975 Nobel Prize in physics for his explanation of the structure of the nucleus.

Louis de Broglie and the Wave Nature of Matter

In 1924, Louis Victor de Broglie (1892–1987) was a graduate student in physics in Paris. De Broglie was intrigued by quantum theory's notion that light has a dual nature. Why not matter? de Broglie asked. If light has a dual nature, should not matter have a dual nature also? Although de Broglie's derivation was more sophisticated than what is being presented here, we can get a feel for what he did by combining Planck's equation for waves from 1900, $E = hc/\lambda$, with Einstein's equation for particles from 1905, $E = mc^2$. Substituting v (velocity) for c (the speed of light) since we are dealing with matter, we can set the two equations equal to each other, as shown:

$$E = hv/\lambda = mv^2.$$

Solving for λ, and cancelling out one term of v, yields the following result:

$$\lambda = h/mv, \text{ or } \lambda = h/p,$$

where p is momentum and is equal to mass times velocity.

De Broglie interpreted this result to mean that a particle traveling with a velocity v has a wavelength associated with it. In our everyday macroscopic world, the wavelength is so short (much, much shorter than the distance across a single atom) that it does not manifest itself. But for an electron confined to the dimensions of an atom, the electron's wavelength is on the same order of magnitude as the size of the atom itself, meaning that the wave nature of the electron is imminently important. This theory allowed de Broglie's to assert a fundamental symmetry to nature: matter and light both have a dual nature in the sense that they have characteristics of both particles and waves (figure 17.2).

Figure 17.2: The Modern Picture of Matter and Light
Since the mid-1920s, scientists have believed that both matter and light possess a dual nature—they both behave as particles and as waves.

	Particles	Waves
Matter	X	X
Light	X	X

Historical Note

The 1929 Nobel Prize in physics went to de Broglie for his discovery of the wave nature of matter, a discovery that set the stage for the next development in physics—the quantum mechanical model of the atom. Dying at the age of ninety-four in 1987, de Broglie lived a long and productive life.

The Quantum Mechanical Model of the Atom

We have arrived at the apex of our story—the quantum mechanical, or wave mechanical, model of the atom. In 1926, the German physicists Werner Heisenberg (1901–76) and Erwin Schrödinger (1887–1961) independently published a model of the atom that was based on the wave nature of matter that had been developed by de Broglie. The mathematics of the quantum mechanical model is very sophisticated. The results are presented here without attempting to justify where they come from. The results were derived rigorously by Heisenberg and Schrödinger for the hydrogen atom, and then extended as approximate results to the remaining elements in the periodic table.

In the remainder of this chapter, the assumption is that we are talking about multi-electron atoms. What is important is not to picture an electron as a particle revolving in an orbit around the atom's nucleus like a planet revolving around the Sun. Instead, the electron is "smeared out" through the region defined by the atom. The electron is a three-dimensional wave with properties that are described by quantum numbers.

Historical Note

Although Heisenberg and Schrödinger developed completely equivalent quantum mechanical models of the atom, we mostly remember Schrödinger for his work. Heisenberg used mathematics that was relatively unfamiliar to physicists. Schrödinger used differential equations, which physicists had been accustomed to using since Newton. Therefore, modern day courses in quantum mechanics revolve around teaching students how to solve the Schrödinger equation.

Quantum Numbers

The description of the electronic structure of an atom begins with four numbers called quantum numbers. The four quantum numbers are described as follows:

➤ The principal quantum number n: The number *n* can take on whole-number values ranging equal to 1, 2, 3, 4, 5, . . . ∞. The value of n is the main factor that determines the energy of an electron. Electrons are grouped in spherically symmetrical regions of space called shells. The value of n specifies the shell in which an electron is located. The shell for n = 1 is closest to the nucleus. As the value of n increases, the corresponding shell becomes larger in size, the energy of the electron increases, and the electron is located farther from the nucleus. There also are letter designations for the shells: n = 1 refers to the K shell, n = 2 to the L shell, n = 3 to the M shell, and so forth in alphabetical order.

➤ The angular momentum quantum number *l*. Just as the value of n designates a shell, the value of *l* designates a subshell within a given shell. For a given value of n, there are n values of *l* that assume the values 0, 1, 2, 3, . . . , n–1. Physicists use the letter *l* (or L) to specify angular momentum, so the value of *l* specifies an electron's angular momentum. In simple terms, angular momentum is a property possessed by a particle that is moving in a circle of radius r. Angular momentum is the product of the particle's mass times its velocity times the radius of revolution. Higher values of *l* mean higher values of angular momentum. There are letter designations associated with values of *l*, just as there are letter designations associated with values of n:

➤ *l* = 0 is an s subshell

➤ *l* = 1 is a p subshell

➤ *l* = 2 is a d subshell

➤ *l* = 3 is an f subshell

➤ *l* = 4 is a g subshell, and so forth.

Because there are n values of *l* for a given value of n, this means there are n subshells in a given shell. Here is an example of how we can list shells and subshells as follows:

➤ K shell n = 1   *l* = 0 (an s subshell)

➤ L shell n = 2   *l* = 0 (s) and 1 (a p subshell)

➤ M shell n = 3   *l* = 0 (s), 1 (p), and 2 (a d subshell)

➤ N shell n = 4   *l* = 0 (s), 1 (p), 2 (d), and 3 (an f subshell)

➤ The magnetic quantum number m_l. A moving electron can be thought of as possessing an electrical current. Magnetic fields are generated by electrical currents, so the value of an electron's angular momentum determines the magnetism associated with the electron. For a given value of the angular momentum quantum number l, there are $2l + 1$ values of m_l that take on the values $-l$, $-l + 1$, $-l + 2, \ldots, -1, 0, +1, \ldots, +l$. The values of m_l designate orbitals, regions of space occupied by electrons with shapes that are determined by the wave properties possessed by the electrons. (The term *orbital* is derived from the term *orbit*. Orbits, however are only two-dimensional. The term *orbital* is used to emphasize that it is referring to something that is three-dimensional.) Thus, in a subshell designated by the quantum number l, there are $2l + 1$ orbitals. Here are some examples:

➤ $l = 0$ s subshell $m_l = 0$ (an s orbital)

➤ $l = 1$ p subshell $m_l = -1, 0, +1$ (three p orbitals)

➤ $l = 2$ d subshell $m_l = -2, -1, 0, +1, +2$ (five d orbitals)

➤ $l = 3$ f subshell $m_l = -3, -2, -1, 0, +1, +2, +3$ (seven f orbitals)

➤ The spin quantum number m_s. There are two magnetic effects associated with an electron in an atom. One effect is due to the angular momentum possessed by the electron. The other effect is an intrinsic magnetism due to what is called the spin of the electron.

Important Point

Each electron in an atom possesses a set of four quantum numbers, designated n, l, m_l, and m_s. Two electrons cannot possess the same set of quantum numbers at the same time. An electron can, however, change the values of its quantum numbers by making transitions to different energy levels.

Spin is actually an unfortunate term. It makes an electron—or any other elementary particle like a proton, which also possesses spin—sound like a spinning top, which, of course, an electron is not. A top is a classical particle description, but an electron confined to an atom behaves as a wave. An electron's property of spin is a purely quantum mechanical property of the electron that has no classical analogy in the everyday macroscopic world in which we live. Therefore, physicists are at a loss to call spin anything else. An electron can exist in one of two possible spin states: up and down. Spin up is assigned a numerical value of +1/2 and spin down is assigned a value of –1/2. Although this is a classical description and not a quantum mechanical one, picturing it as a spinning ball is the best we can do.

The Relationship between Quantum Numbers and the Arrangement of Electrons

We can summarize what has been said so far about quantum numbers by stating the following:

➤ Electrons are arranged in energy levels, or shells, which are designated by the principal quantum number n. Low values of n mean low energies and shells that are close to the nucleus. High values of n mean high energies and shells that are farther from the nucleus.

➤ Shells are subdivided into subshells, which are designated by the angular momentum quantum number l. In a shell designated by quantum number n, there are n subshells labeled s, p, d, f, etc.

➤ Subshells are further subdivided into orbitals, which are designated by the magnetic quantum number m_l. In a subshell designated by quantum number l, there are $2l + 1$ orbitals.

➤ Orbitals are designated s, p, d, f, etc., according to the subshell in which they are located.

➤ Each electron has a fourth quantum number m_s, called its spin. Spin can take on two values: +1/2 and –1/2, or up and down.

Historical Note

The discovery of the Exclusion Principle won Wolfgang Pauli the 1945 Nobel Prize in physics.

The Pauli Exclusion Principle

A fundamental principle of quantum theory is the Pauli Exclusion Principle, which was formulated by the German physicist Wolfgang Pauli (1900–58), which states that no two electrons in the same atom can have the same set of quantum numbers. Because an individual orbital is specified by values of the first three quantum numbers—n, l, and m_l—the effect of the Pauli Exclusion Principle is that an individual orbital can be occupied by at most two electrons, one with spin up and one with spin down.

Why No More than Two Electrons Can Occupy the Same Orbital

If two electrons in the same orbital were either both spin up or both spin down, then all four of their quantum numbers would be the same, which would violate the Pauli Exclusion Principle. We can offer a qualitative explanation for why this must be so. Remember that electrons are all negatively charged, so they tend to repel each other. An orbital defines an extremely tiny volume of space, so the repulsive force between two electrons in a single orbital is quite large.

The spin of an electron makes the electron behave like a little magnet. We can think of spin up as a bar magnet with its North Pole pointing upward, and spin down like a bar magnet with its North Pole pointing downward. If we try to bring together the north poles of two bar magnets, the magnets repel each other. However, if we bring together the North Pole of one magnet and the South Pole of the other magnet, the magnets attract each other. In the same way, two electrons that are both spin "up" (we say the spins are parallel, or *unpaired*) repel each other, but two electrons with opposite spins (we say the spins are antiparallel, or paired) attract each other. The magnetic attraction between electrons with paired spins tends to counterbalance the electrostatic repulsion between the electrons, allowing the two electrons to share the same orbital.

Important Point

The concept of spin tends to emphasize the particle nature of the electrons, yet the name wave mechanical model is meant to emphasize the wave nature of the electrons. This is an unresolved dilemma in quantum physics. With no analog of electron spin in our everyday macroscopic world, any attempt we make to try to explain spin will be inadequate.

The Aufbau Principle

In addition to the Pauli Exclusion Principle, there are two other principles used to draw an energy level diagram. The first is called the Aufbau Principle, where the German word *aufbau* means "building up." The Aufbau Principle says that we build up the energy levels of an atom by starting with the lowest-energy subshells and working up in energy. Electrons are placed in a subshell until that subshell is filled, and then electrons begin filling in the next highest energy subshell.

Hund's Rule

The second principle is called Hund's Rule, named for its developer, German physicist Friedrich Hund (1896–1997). Hund's Rule is sometimes called the rule of maximum multiplicity. It says that electrons occupying the same subshell tend to have the same spin. The result is to spread the electrons over as many orbitals in the subshell as possible before filling orbitals.

Historical Note

Of all the scientists mentioned in this book, Friedrich Hund lived one of the longest lives. Born in 1896, Hund died in 1997 at the age of 101.

There are two ways to understand Hund's Rule. The first is an argument based on symmetry. Consider a p subshell for example. If the subshell contains three electrons, it is more symmetrical to place one electron in each orbital than to have two electrons in the first orbital, a single electron in the second orbital, and the third orbital unoccupied. In atoms and molecules, symmetrical configurations of electrons tend to have lower energy and thus are preferred over higher energy, asymmetrical configurations.

The second argument follows from the electrostatic repulsion between electrons. It seems more reasonable to separate the electrons in a subshell as much as possible before squeezing two electrons together into a single orbital. Having the electrons' spins parallel also helps keep the electrons apart.

Electronic Configurations

Table 16.1 shows the arrangement of shells, subshells, and orbitals in an atom. In the designation of an orbital, 1s means n = 1 and l = 0 (an s subshell), 2p means n = 2 and l = 1 (a p subshell), and so forth. No more than two electrons can occupy any single orbital. An s subshell has only a single orbital, so its occupancy is also limited to two electrons. However, since a p subshell consists of three orbitals, a maximum of six electrons can occupy a p subshell. A d subshell consists of five orbitals, so it can be occupied by ten electrons. An f subshell has seven orbitals, so it can be occupied by fourteen electrons. In general, for a given value of l, there are $2l + 1$ orbitals, and 2 times $2l + 1$, or $4l + 2$, possible electrons.

Table 17.1 – How Shells, Subshells, and Orbitals Are Arranged

n	*l*	Orbital	m_l	Maximum Number of Electrons Subshell	Shell
1	0	1s	0	2	2
2	0	2s	0	2	
2	1	2p	−1, 0, +1	6	8
3	0	3s	0	2	
3	1	3p	−1, 0, +1	6	
3	2	3d	−2, −1, 0, +1, +2	10	18
4	0	4s	0	2	
4	1	4p	−1, 0, +1	6	
4	2	4d	−2, −1, 0, +1, +2	10	
4	3	4f	3, −2, −1, 0, +1, +2, +3	14	32

Since the nth energy level contains n subshells, it follows that the K shell can be occupied by a maximum of two electrons, the L shell by eight electrons, the M shell by eighteen electrons, the L shell by thirty-two electrons, and so forth. In general, an energy level with a given value of n can be occupied by a maximum of $2n^2$ electrons.

Scientists have a shorthand notation to indicate how many electrons occupy each orbital of an atom. This notation is called the electronic configuration of an element. Unless stated otherwise, it is assumed that the configuration represents the ground state of the atom, i.e., that all of the elements are in the lowest-energy subshells available. Superscripts indicate the number of electrons in a subshell. Table 17.2 gives the electronic configurations for the first eleven elements.

Table 17.2 – Electronic Configurations of the First 11 Elements

Atomic Number	Element	Electronic Configuration
1	H	$1s^1$
2	He	$1s^2$
3	Li	$1s^2 2s^1$
4	Be	$1s^2 2s^2$
5	B	$1s^2 2s^2 2p^1$
6	C	$1s^2 2s^2 2p^2$
7	N	$1s^2 2s^2 2p^3$
8	O	$1s^2 2s^2 2p^4$
9	F	$1s^2 2s^2 2p^5$
10	Ne	$1s^2 2s^2 2p^6$
11	Na	$1s^2 2s^2 2p^6 3s^1$

Energy Level Diagrams

A graphical representation of the energy levels, subshells, and orbitals occupied by the electrons in an atom is called an energy level diagram. Unlike electronic configurations, which only show the occupancy of subshells, energy level diagrams show the occupancy of individual orbitals. Electrons are represented using up and down arrows to indicate the electrons' spin.

Conclusion

In the next chapter we will take a look at the properties of families of elements in the periodic table. Mendeleev died before anyone could explain why his periodic table has the form that it does. The quantum mechanical model of the atom, however, explains the structure of the periodic table.

CHAPTER 18

The Periodic Properties of Elements

In This Chapter

- ➤ Families of elements
- ➤ Sizes of atoms
- ➤ Ionization energies
- ➤ Electron affinities
- ➤ Electronegativity
- ➤ Melting points of elements

Given what we have now learned about electronic configurations of atoms, we can explain the structure of the periodic table. To put it quite simply, elements occupying the same column of the periodic table have similar chemical and physical properties because they have similar electronic configurations. Properties are determined largely by the kinds and numbers of valence electrons—the electrons occupying the highest energy levels. To understand this, let us look at a few families of elements.

Families of Elements

Some of the most common elements found on Earth occupy positions in the first two columns of the periodic table. The elements in the first column, the alkali metals, exhibit regular trends in chemical and physical properties that are apparent as the size of their atoms increases. Elements in the second column, the alkaline earths, also exhibit similar trends. Observations about trends in these two families can be generalized to understanding the properties of elements belonging to other families in the periodic table.

The Alkali Metals

The alkali metals occupy the first column of the periodic table. Their electronic configurations are the following:

> ➤ Lithium: $1s^2 2s^1$

> ➤ Sodium: $1s^2 2s^2 2p^6 3s^1$

> ➤ Potassium: $1s^2 2s^2 2p^6 3s^2 3p^6 4s^1$

> ➤ Rubidium: $1s^2 2s^2 2p^6 3s^2 3p^6 3d^{10} 4s^2 4p^6 5s^1$

> ➤ Cesium: $1s^2 2s^2 2p^6 3s^2 3p^6 3d^{10} 4s^2 4p^6 4d^{10} 5s^2 5p^6 6s^1$

Chem Vocab

Isoelectronic means "having the same number of electrons," or "having the same electronic configuration."

Important Point

We would never find alkali metals in nature as the neutral elements because they are extremely reactive. We only find alkali metals as +1 ions, for example, in rock salt—NaCl.

What we see is that each alkali metal has a single outermost (valence) electron in an s subshell. That commonality gives each element similar properties. For example, they are all extremely reactive metals. Each alkali metal reacts readily, sometimes even explosively, with water. Therefore, none of them is found in nature in pure form. All of these elements form +1 ions—Li^+, Na^+, K^+, etc. The reason for that is that by losing one electron, alkali metals become isoelectronic (having the same number of electrons) as the noble gas elements in the preceding rows. Therefore, like the noble gases, alkali metals are extremely stable entities.

The only compounds alkali metals form are ionic compounds such as NaCl, $NaNO_3$, KI, and K_2CO_3. These compounds all tend to be soluble in water. They tend to be fairly soft metals with a low melting point (for metals). Like metals in general, they conduct heat and electricity.

The Alkaline Earths

The alkaline earths occupy the second column of the periodic table. Their electronic configurations are the following:

➤ Beryllium: $1s^22s^2$

➤ Magnesium: $1s^22s^22p^63s^2$

➤ Calcium: $1s^22s^22p^63s^23p^64s^2$

➤ Strontium: $1s^22s^22p^63s^23p^63d^{10}4s^24p^65s^2$

➤ Barium: $1s^22s^22p^63s^23p^63d^{10}4s^24p^64d^{10}5s^25p^66s^2$

Each alkaline earth element has two outermost (valence) electrons in an s subshell. Alkaline earths are all reactive metals, although less so than the alkali metals are. Like the alkali metals, none of the alkaline earths is found in nature in pure form. All of these elements form +2 ions—Be^{2+}, Mg^{2+}, Ca^{2+}, etc.—because that makes them isoelectronic with the nearest noble gas elements. With the exception of beryllium, they only form ionic compounds such as $MgCl_2$, $CaCO_3$, and $Ca(NO_4)_2$.

Important Point

Beryllium's chemical properties are actually more like those of aluminum than they are of the other alkaline earths. Beryllium and aluminum both tend to exhibit covalent bonding, rather than ionic bonding.

Alkaline earth compounds are often insoluble in water. Like metals in general, they conduct heat and electricity.

Trends in Sizes of Atoms in the Periodic Table

Two trends in the sizes of atoms exist:

➤ With a few exceptions, atoms tend to get larger as one descends a column (group or family) of the periodic table

➤ Again with a few exceptions, atoms tend to get smaller as one moves across a row (period) of the periodic table from left to right.

Both of these trends are fairly easy to understand. Recall that electron shells increase in size as the value of the quantum number n increases. Look at the first family of elements, the alkali metals. Lithium's outermost electron is in a "2s" subshell, sodium's outermost electron is in a 3s subshell, potassium's is in a 4s subshell, and so forth. Since each element's outermost electron is in a subshell with a larger value of n, each shell is larger in size, and thus each atom is larger in size.

On the other hand, as we move across a period from left to right, additional electrons are being added to the same energy level, i.e., the value of n does not change. Look at the second period of elements, lithium to neon. An atom of lithium has three protons in its nucleus, beryllium has four protons, boron has five protons, and so forth until we reach neon with ten protons in its nucleus. All the protons have positive charges. As we increase the positive charge on the nucleus, the total electrostatic force on all the electrons increases. All of the electrons are pulled a little closer to the nucleus. Therefore, as the charges on the nuclei increase, the atoms become a little smaller in size.

Important Point

A good rule of thumb is to think of metal atoms as being relatively large and nonmetal atoms as being relatively small. Cesium has the largest atoms of naturally occurring elements. Hydrogen and helium have the smallest atoms.

Trends in Sizes of Ions

Metals give up electrons to become positive ions. Consider sodium as an example. The electronic configuration of a sodium atom is $1s^2 2s^2 2p^6 3s^1$. A sodium atom (Na) has an electron in the third energy level. When a neutral atom of sodium gives up an electron, it is the 3s electron—sodium's outermost electron—that it loses. The resulting electronic configuration of a sodium ion (Na$^+$), therefore, is $1s^2 2s^2 2p^6$. A sodium ion's outermost electrons are only in the second energy level. The result is that a sodium ion is smaller than a sodium atom. All simple positive ions (ions composed of only one atom) are smaller than the neutral atoms from which they are derived.

Nonmetals

Conversely, nonmetals gain electrons to become negative ions. Consider fluorine as an example. The electronic configuration of a fluorine atom (F) is $1s^2 2s^2 2p^5$. Gaining one electron makes the electronic configuration of a fluoride ion (F$^-$) $1s^2 2s^2 2p^6$. Since the fluoride ion's additional electron is in the same energy level as the other valence electrons, it would be easy to conclude that there should be no change in size. Remember, however, that all of the electrons carry negative charges. They all repel each other. That mutual repulsion among a fluoride ion's six 2p electrons causes them all to move farther apart from each other. The result is that a fluoride ion is larger than a fluorine atom. All simple negative ions are larger than the neutral atoms from which they are derived. When thinking about elements in general, and not specific elements, a good rule of thumb is that metals tend to have small ions and nonmetals tend to have large ions.

Important Point

Metal ions are smaller than their neutral atoms. Nonmetal ions are larger than their neutral atoms.

Trends in Ionization Energies

Ionization is the process by which an atom loses an electron to become a positive ion. We can write that as the reaction $M \rightarrow M^+ + e$. An atom must absorb energy in order for an electron to be removed. That energy is called the ionization energy of that element. The energy necessary to remove the first electron is called the first ionization energy. The energy necessary to remove the second electron is called the second ionization energy, and so forth. Ionization energies are often measured in the gas phase so that only single atoms are involved.

Chem Vocab

Ionization energy is the amount of energy that an atom must absorb for an electron to be removed from the atom.

It should make sense that the farther an electron is from an atom's nucleus, the less the electrostatic force of attraction will be between that electron and the nucleus with its positively charged protons. In addition, in the case of atoms with many electrons, the inner electrons—the electrons that are located in shells that are closer to the nucleus—will tend to at least partially shield electrons that are farther away from the nucleus so that they do not feel the full positive charge of the nucleus. The result is that as the size of atoms increases, their ionization energy decreases. Since the size of atoms increases descending a column of the periodic table, ionization energy decreases descending a column.

On the other hand, since size of atoms decreases when moving across a row of the periodic table from left to right, the attraction an electron feels to its atom's nucleus increases. Therefore, as we move across a row of the periodic table from left to right, ionization energy increases.

When thinking about elements in general, a good rule of thumb is that metals tend to have low ionization energy and nonmetals tend to have high ionization energy. Therefore, we may think of cesium as being the element with the lowest ionization energy and helium as being the element with the highest ionization energy. This difference between metals and nonmetals results from the differences in size of atoms between metals and nonmetals and explains why metals are conductors of electricity and nonmetals are nonconductors.

Important Point

Metals have low ionization energy; nonmetals have high ionization energy.

Chem Vocab

Electron affinity is defined as the amount of energy required to remove an electron from a negative ion that is in the gaseous state.

Important Point

Metals have low electron affinities. Nonmetals have high electron affinities.

Large atoms have low ionization energy, which means that their outermost one or two (or sometimes three) electrons are easily removed. Applying a small electrical potential difference across a metal (by connecting a battery to it, for example) supplies enough energy to cause those outermost electrons to move to other atoms. The flow of electrons through a metal is what we mean when we say that a metal conducts an electrical current. Conversely, the high ionization energy of nonmetals means that nonmetal atoms hold tightly onto their electrons. It is very difficult to get electrons to flow, so nonmetals do not conduct electrical currents.

Trends in Electron Affinities

Electron affinity is defined as the amount of energy required to remove an electron from a negative ion that is in the gaseous state. Since metals tend not to form negative ions, electron affinity really only applies to nonmetals. The trends for electron affinity are similar to the trend for ionization energy. As we move across a row of the periodic table from left to right, electron affinity generally increases. As we descend a column, electron affinity tend to decrease.

Trends in Electronegativities

Electronegativity is defined as the tendency of an atom that is bonded to another atom to attract the bonding electrons. As a general rule, nonmetals are more electronegative than metals are. Values of electronegativity are usually expressed on a scale of 0 to 4, with 4 being the most electronegative. The noble gases have electronegativities of 0. (There are no units for electronegativities; the numbers just indicate relative tendencies.) Elements with the highest values of electronegativity are the nonmetals with the exception of the noble gases, which are assigned a value of 0 because they tend not to form chemical bonds. Fluorine is the most electronegative element in the periodic table and is assigned an electronegativity of 4.0.

Chem Vocab

Electronegativity is defined as the tendency of an atom that is bonded to another atom to attract the bonding electrons.

Elements close to fluorine also have relatively high electronegativities (3+), and electronegativities decrease with distance from fluorine in the table. Because the location of hydrogen in the periodic table can be considered to be both in the alkali metal column (I) and in the halogen column (VII), hydrogen's electronegativity is a special case and lies between the electronegativities of boron and carbon.

Generally speaking, electronegativities increase as we move across a row of the periodic table from left to right and as we ascend a column of the table. The result is that the closer an element is to fluorine in the periodic table, the more electronegative the element is. When two elements are the same distance from fluorine, the element with smaller atoms is the more electronegative. Comparing oxygen and chlorine, for example, oxygen is one square to the left of fluorine and chlorine is one square below fluorine. Oxygen atoms are smaller than chlorine atoms, so oxygen is more electronegative than chlorine.

Important Point

Fluorine is the most electronegative element in the periodic table. That fact correlates with F_2 being an extremely powerful oxidizing agent—the tendency of F_2 to gain electrons makes it do so violently in many cases.

Comparisons of electronegativity most often occur with discussions of polar covalent bonds (see Chapter 19). In a nonpolar covalent bond, two atoms share two or more electrons equally. In a polar covalent bond, the electrons are shared unequally. There is a slight concentration of negative charge on the more electronegative atom, leaving a slight concentration of positive charge on the less electronegative atom of the pair.

Chem Vocab

In a nonpolar covalent bond, the bonding electrons are shared equally by the atoms forming the bond.

In a polar covalent bond, the bonding electrons are shared unequally. The bonding electrons, on average, will be closer to the more electronegative of the two atoms forming the bond.

Trends in Melting Points of Metals

As a general rule, metals on the left side of the periodic table and metals on the right side tend to have a lower melting point than metals located toward the middle of the table. Thus, alkali metals (with a melting point from 29°C to 181°C [84°F to 358°F]), alkaline earths (700°C to 839°C [1292°F to 1542°F]), and post-transition metals (30°C to 328°C [86°F to 622°F]) generally melt at a relatively low temperature for metals. (Exceptions are beryllium and aluminum, which have properties that are more like each other than like the other metals in their families.) Two familiar post-transition metals are tin and lead. Their low melting point is taken advantage of in the use of tin in solder and in the use of lead in bullets.

The metal with one of the highest melting points is tungsten at 3,014°C (5,457°F). Tungsten's exceptionally high melting point was the reason that the American Thomas Edison (1847–1931) finally settled on tungsten as the filament for his incandescent lightbulb. It should be noted that rhenium has a slightly higher melting point than tungsten, but rhenium is too expensive to use in lightbulbs.

Important Point

The term *incandescent* means that a substance glows when it is heated. Many metals would melt before they would begin to glow, so that lightbulbs made with them would be short-lived. Tungsten's high melting point keeps tungsten in the solid state as it glows.

Historical Note

Thomas Edison did not actually invent the electric lightbulb. What he did succeed in doing was to invent the first lightbulb that was durable enough to be commercially successful.

Other high-melting metals tend to be located near tungsten—zirconium, hafnium, molybdenum, technetium, rhenium, ruthenium, osmium, rhodium, iridium, and platinum—and have a melting point that ranges from 1,772°C to 3,180°C (3,222°F to 5,756°F).

The metal with the lowest melting point, of course, is mercury, which melts at −39°C (−38°C) and is already a liquid at room temperature. Gallium and cesium also have a low melting point, melting just above room temperature. Gallium is also unique among the metals in that it has the widest range of temperatures for which it exists as a liquid. Gallium melts at 30°C (86°F) and boils (at 1 atm pressure) at 2,205°C (4,000°F), making gallium a liquid over a range of 2,175°C (3,914°F).

Conclusion

As stated earlier, the periodic table of the elements is the single most important unifying principle in chemistry. Chemistry students and chemistry practitioners alike refer to the periodic table constantly as they seek to understand the properties of the elements and the compounds that they form.

PART FOUR

The States of Matter

CHAPTER 19

The Nature of the Chemical Bond

Atoms in molecules, ions in crystals, and atoms in crystals are attracted to each other by electrostatic forces—the forces between positively charged protons and negatively charged electrons. These attractive forces are what chemists call chemical bonds. Electrostatic forces are extremely strong and their energy is stored in bonds.

The Nature of Stored Energy

To understand how energy is stored, let us look at a mechanical analogy. Suppose we lift a heavy box from the floor and set it on a table. To do so we have to exert a force against the opposing force of gravity. That means we have to do work. Work is force times distance. (By Newton's third law of motion, which states that forces come in pairs—whenever a force is pushing something, an equal and opposite force is pushing back. In this case that opposite force is gravity itself.)

Work done against the force of gravity is calculated in this case by multiplying the weight of the box times the height of the table. If weight is in pounds and height is in feet, then the units of work are foot-pounds (ft-lb). In the metric system, weights are measured in newtons and distances are measured in meters. The unit of work is the newton-meter (N-m), which is equivalent to the joule (J). But doing work changes the box's energy. In this case, in the absence of friction, the change in the gravitational potential energy of the box is equal numerically (and is a positive number) to the work that was done (energy is measured in foot-pounds, or joules, also). Gravitational potential energy is called potential energy because it represents energy that is being stored and that can be used to do work, in this case, the work that would be done by gravity if the box were lowered to the floor again.

The Positive and Negative of Electrostatic Forces

When two atoms (call them A and B) come together, the combinations of electrostatic forces are more complex. There actually are four interactions:

> ➤ The protons in the nucleus of atom A feel an attraction to the electrons in atom B

> ➤ The electrons in atom A feel an attraction to the protons in the nucleus of atom B

> ➤ The protons in atom A feel a repulsive force due to the protons in atom B

> ➤ The electrons in atom A feel a repulsive force due to the electrons in atom B

You might think that all of these attractive and repulsive forces would just cancel each other out, but that is not what happens. Because of the way electrons are distributed in an atom (remember that an atom is mostly empty space), and the protons are concentrated in a very tiny nucleus, the attractive forces win over the repulsive forces, and there is a net force of attraction.

Just as in our analogy of work being done when the box was being lifted, work is also done to force neutral atoms to approach each other. In this case, however, the work is not being done by a force outside the atoms (as we were outside the box when we lifted it), but by the force of attraction between the atoms themselves. Just as we were doing work by lifting the box, with atoms the electrostatic force is doing work. In this case the stored energy is called electrostatic potential energy. We might also call this energy chemical energy. To separate the two atoms again, an external force is required to pull them apart, a force that would now do an amount of work equal in magnitude to the stored chemical energy in the bond.

You may be wondering if the net electrostatic force is pulling the two atoms together, do they come so close together that effectively the two nuclei touch each other? The answer is no, they do not. Remember that there are repulsive forces between the electrons on the two atoms and between the protons on the two atoms. Eventually the forces of repulsion will balance the forces of attraction. When that happens, the atoms' two nuclei settle into a fairly stable equilibrium distance from each other. The distance between the two nuclei is called the bond length. Chemists can measure bond lengths and how much energy is stored in chemical bonds.

Chem Vocab

Bond length is the distance between the nuclei of two atoms that are bonded together in a molecule. Bond length varies with the state the substance is in (solid, liquid, or gas) because the distance between nuclei is different in different states.

Ionic Bonds: Transferring Electrons

Ionic bonds tend to form between metals and nonmetals. The metal atoms give up one or more electrons to form positive ions and the nonmetal atoms gain one or more electrons to form negative ions. Common examples of ionic compounds are ones in which the metal is an alkali or alkaline earth element and the nonmetal is oxygen, sulfur, or a halogen.

Consider the ionic compound sodium chloride (NaCl), which consists of a lattice of Na^+ and Cl^- ions that alternate in three dimensions. Even though there are not discreet molecules of NaCl, as there are, for example, with Cl_2, we can isolate one pair of positive and negative ions and call that a molecule.

The positive charge on the Na^+ ion is there because the ion has one less electron than a neutral Na atom does. The negative charge on the Cl^- ion is there because the ion has one more electron than a neutral Cl atom does. Opposite charges attract each other. The attraction between a positive ion and a negative ion is what chemists call the ionic bond. Ionic bonds tend to be very strong. Because there are so many ionic bonds in a macroscopic-sized crystal and they extend in all three dimensions, ionic compounds can be expected to be hard solids at room temperature.

Covalent Bonds: Sharing Electrons

Let us consider the case of atoms of two nonmetals (or of a nonmetal and a metalloid) coming together. Once the chemical bond between the two atoms has been established and the atomic nuclei have settled into their equilibrium distance apart, the outer electron shells on the two atoms will overlap. This results in some of the electrons in the two atoms being shared. Because electrons have up spins and down spins, it is most common for electrons to be shared in pairs, one with spin up and one with spin down.

Important Point

Because covalent bonds are formed by sharing pairs of electrons, three common kinds of bonds result:

> ➤ Single bonds, where two electrons are shared. This is the most common kind of bond.

> ➤ Double bonds, where four electrons are shared.

> ➤ Triple bonds, where six electrons are shared.

Lewis Electron Dot Structures

It is very common in chemistry to draw structures of chemical compounds using dots or dashes to connect atoms together. A single dot represents one electron, and a pair of dots represents a pair of electrons. A dash also represents a pair of electrons. Either way, a pair of electrons connecting two atoms is our usual representation of a chemical bond.

The University of California physical chemist Gilbert Newton Lewis (1875–1946) introduced a notation to represent atoms and their valence (outermost) electrons that today we call Lewis electron dot structures in his honor. Here are the dot structures for first ten elements in the periodic table:

H·   He: Li·   ·Be·   ·B·   ·C·   ·N·   :O·   :F·   :Ne:

Elements in the same column as one of the above elements have analogous electron dot structures. For example, silicon's structure is ·Si·. An important feature of Lewis's symbols for atoms is that the number of unpaired electrons in the symbol is usually the number of chemical bonds that element will form. Thus, a hydrogen atom can form only one bond. An oxygen atom can form two bonds, a nitrogen atom three bonds, and a carbon atom four bonds.

Chem Vocab

Valence electrons are the electrons in the outermost energy level of an atom. For example, the electronic configuration of a ground-state lithium atom is $1s^2 2s^1$. Since there is one electron in the second energy level, lithium atoms have one valence electron. The electronic configuration for a ground-state oxygen atom is $1s^2 2s^6$, so oxygen atoms have six valence electrons.

We can combine Lewis structures for atoms to make Lewis structures for molecules. For example, water would be represented as $\ddot{\text{O}}$:H. Methane would be represented as H:$\ddot{\text{C}}$:H. We will return to Lewis structures in the next chapter when we look at the structures of molecules.

Electronegativity

Each element in the periodic table has its own tendency when it is part of a molecule to pull the electrons being shared closer to its nucleus than to the nucleus of the other atoms in the molecule. This tendency to pull electrons toward itself is called the element's electronegativity.

Polar Covalent Bonds

If atoms of different elements (C and N, for example) form a chemical bond, the more electronegative of the two elements will exert a net pull on the bonding electrons. This will result in a separation between the center of positive charge (due to the protons) and the center of negative charge (due to electrons). Such a situation creates what physicists call an electric dipole. A dipole exists when two charges, q_1 and q_2, which are equal in magnitude but opposite in sign (one is + and the other is −) are separated by a distance d. The strength of the dipole (designated by the Greek letter mu, μ) is calculated by the expression $\mu = q \cdot d$, where μ is a positive number with units of Debyes (D). If two atoms of the same element (H_2, for example), however, form a chemical bond, then the two atoms have the same electronegativity and there is no dipole; the electrons are shared equally.

Molecules with permanent dipoles are said to be polar. Molecules without permanent dipoles are said to be nonpolar. One of the most important properties a molecule has is whether it is polar or nonpolar. Water (H_2O) is a familiar example of a polar molecule. The polarity of a water molecule determines many of water's anomalous properties (see Chapter 24).

Important Point

Whether a molecule is polar or nonpolar is one of the most important properties a molecule has.

Nonpolar Covalent Bonds

When bonding electrons are shared equally, then the bond is said to be a nonpolar covalent bond. This only happens when the bond is formed between atoms of the same element. In the case of diatomic molecules, there are several examples: H_2, N_2, O_2, F_2, Cl_2, Br_2, and

I_2. H_2 and the halogens all have single bonds, which can be represented by either two dots (representing the two electrons)—H:H, for example—or a single line connecting the atoms, as in H–H. In the oxygen molecule, O_2, the two atoms are connected by a double bond (four electrons). O_2 can be represented as O::O or O=O. Nitrogen molecules, N_2, are connected by a triple bond, represented by N:::N or N≡N. Larger molecules that consist of only one element include O_3, P_4, and S_8.

The more electrons that are being shared, the stronger the bond is. Thus, double bonds are stronger than single bonds, and, in turn, triple bonds are stronger than double bonds. A stronger bond means that more energy is required to break the bond and pull the atoms apart. The requirement of more energy means the bond is less chemically reactive.

The relative strength of chemical bonds helps explain the relative abundance of gases in the atmosphere, of which about 99 percent of air is N_2 and O_2. Since H_2 has only a single bond, H_2 molecules are rare in the atmosphere. Hydrogen is very reactive and the atoms quickly combine with other elements to form water or other molecules. The oxygen molecule's double bond is much stronger, so O_2 is less reactive. That is why the atmosphere is about one-fifth O_2; oxygen can accumulate. The bond cannot be too strong, though, or living organisms would be unable to break that bond to use oxygen in the essential process of respiration. Of course, if oxygen's bond were any weaker, fires would burn so fiercely that it would be extremely difficult to extinguish them.

Finally, the much greater strength of nitrogen's triple bond is responsible for the fact that the atmosphere is about four-fifths N_2. Nitrogen gas is very unreactive. Although nitrogen is an essential element in all living organisms—being a component of both proteins and nucleic acids—it is very difficult for living organisms to access atmospheric nitrogen. Animals cannot do it at all. Every breath we take has four times as many nitrogen molecules in it as oxygen molecules, but all we can do with the nitrogen is breathe it right back out again—our lungs have no way of absorbing it. Even plants cannot use nitrogen from the atmosphere directly. Microorganisms first must convert N_2 into ammonia (NH_3) and other usable forms of nitrogen before plants can use it. In turn, animals (including us!) obtain 100 percent of the nitrogen in their bodies from the plants they eat.

Conclusion

You might be wondering if there are any kinds of chemical bonds in which the number of electrons shared is something other than 2, 4, or 6. The answer is yes. A single electron can be shared, in which case the bond is called a half bond. Some of the transition metals can form quadruple bonds, in which eight electrons are shared. However, those kinds of bonds tend to be beyond the level of this book and are covered in more advanced treatises on inorganic chemistry.

CHAPTER 20

Shapes of Molecules

In This Chapter

➤ More Lewis electron dot structures

➤ Shapes of molecules

➤ Polarity

➤ Resonance

A fundamental principle of chemistry: An understanding of the structure of matter at the atomic and molecular level is the key to understanding the properties of materials at the everyday, macroscopic level. We have already considered the structure of atoms (Chapters 15 and 16) and the nature of the chemical bonds that hold atoms together within molecules (Chapter 19). In this chapter we will look at the structure of molecules.

Important Point

The structure of matter at a microscopic level determines the properties of matter at a macroscopic level.

Lewis Electron Dot Structures for Molecules

We saw in Chapter 19 how to draw Lewis electron dot structures for atoms and how to combine them to make diatomic molecules. In this chapter we will do the same with molecules that contain three or more atoms. To do so, we will follow two fundamental rules:

➤ Beginning with the element carbon, each nonmetal atom in a molecule will be represented as being surrounded by eight valence electrons. This is called the octet rule. The reason for the octet rule is that having eight valence electrons makes a nonmetal atom isoelectronic with the noble gas at the end of its period, which gives that atom considerable stability.

➤ The number of valence electrons represented in a Lewis structure for a molecule must equal the sum of the numbers of valence electrons in the atoms that make up that molecule. For example, when drawing the Lewis structure for CO_2, we observe that carbon has four valence electrons and each oxygen has six valence electrons. Since $4 + 6 + 6 = 16$, the Lewis structure for CO_2 must have 16 electrons.

Although there are molecules that do contain an odd number of electrons, in this discussion we will limit our examples only to molecules that contain an even number of electrons. Therefore, in the Lewis structure of a molecule, all of the electrons will be shown as being paired.

Table 20.1 shows some of the shapes that molecules can have.

Table 20.1: Lewis Structures of Common Molecules

Methane: CH_4 (8 electrons)	H:C̈:H (with H above and below)
Ammonia: NH_3 (8 electrons)	H:N̈:H (with H below)
Water: H_2O (8 electrons)	H:Ö:H
Carbon dioxide, CO_2 (16 electrons)	Ö::C::Ö
Sulfur dioxide: SO_2 (18 electrons)	:Ö:S̈::Ö
Ethyl alcohol: C_2H_5OH (20 electrons)	H:C:C:Ö:H (with H H above and H H below)

Valence Shell Electron Pair Repulsion Theory

The phrase *valence shell electron pair repulsion*, or VSEPR, sounds like a mouthful, but it is actually fairly simple to understand. *Valence shell* means we are only considering the valence electrons of an atom; *electron pair* means that in a molecule, the electrons are always paired. *Repulsion* is the key term. Electrons are all negatively charged. VSEPR states that the electron pairs will move into relative orientation so as to maximize the distance the electron pairs are from each other.

Important Point

The atoms in molecules move into relative positions to maximize the distances between their valence electrons because that arrangement gives the configuration of atoms with the lowest possible energy.

Consider each example in Table 20.1. In the case of methane (CH_4), there are four pairs of electrons around the carbon atom (necessarily so in order to satisfy the octet rule). What geometrical arrangement can those four pairs of electrons assume that will maximize their distance apart from each other?

At first, you might be tempted to say that the pairs of electrons will assume positions at the four corners of a square. In a square, however, the angles between bonds would be only 90°. In addition, a square is flat, and many molecules are three-dimensional. Therefore, a better answer to the question is that the four pairs of electrons can assume positions at the vertices of a tetrahedron.

In a tetrahedron the angles between bonds are 109.5°, which is greater than 90°, which means that the pairs of electrons are further apart in a tetrahedral arrangement than they would be in a square. Since four hydrogen atoms are attached to the carbon atom, they also will occupy the vertices of a tetrahedron. The bonding in methane is best described as there being four single bonds around the carbon atom, and that the "shape" of a methane molecule is tetrahedral.

Important Point

The most common shapes of molecules in nature are tetrahedral, pyramidal, bent, triangular, and linear.

In the case of ammonia (NH_3), there are still four pairs of electrons around the nitrogen atom, but there are only three atoms of hydrogen attached to the nitrogen. The geometrical shape of the electron pairs is still tetrahedral, but the shape of the NH_3 molecule is now said to be pyramidal. *Pyramidal* means "like a pyramid," but not exactly like the pyramids in Egypt, since Egyptian pyramids have square bases, not triangular ones. We say that the bonding in ammonia is best described as there being three single bonds and one nonbonding pair of electrons around the nitrogen atom, and that the shape of an ammonia molecule is pyramidal.

The angle between the bonds in NH_3 is a little bit less than 109°. This time the angle is about 107°. The reason for this is that the region of space occupied by the nonbonding electrons is larger than the regions of space occupied by the bonding pairs of electrons. Therefore, the nonbonding electrons are able to push the bonding pairs of electrons a little closer together than they otherwise would have, resulting in an angle between bonds that is less than 109°.

In the case of water (H_2O), there are still four pairs of electrons around the oxygen atom, but now there are only two atoms of hydrogen attached to the oxygen, also giving oxygen two nonbonding pairs of electrons. As before, the geometrical shape of the electron pairs is tetrahedral, but the shape of the H_2O molecule is said to be bent.

The bonding in water is best described as there being two single bonds and two nonbonding pairs of electrons around the oxygen atom. Notice that the angle between bonds is even smaller than it is in ammonia—about 104°. This is because there are now two nonbonding pairs of electrons pushing on the bonding pairs of electrons, causing the bonding pairs to be even closer together than they were in either methane or ammonia.

So far our examples have only had single bonds between atoms. Looking next at carbon dioxide, what would happen if we tried to draw CO_2 with only single bonds. CO_2 would come out looking like this: $:\ddot{O} - \ddot{C} - \ddot{O}:$, with twenty valence electrons in the picture instead of the sixteen to which we are limited. To reduce the number of valence electrons, we need to write the carbon-oxygen bonds as double bonds, at the same time still satisfying the octet rule around each atom. That means that there are electrons on only two sides of the carbon atom. Using VSEPR, we conclude that the geometrical arrangement that maximizes the separation between those two sets of electrons is a straight line with a 180° angle between bonds. We describe the bonding around the carbon atom as two sets of double bonds and the shape of the molecule as linear.

In SO_2 we now encounter a mix of double and single bonds. Sulfur and oxygen atoms both have six valence electrons, so that SO_2 has a total of eighteen valence electrons. The only way we can draw a Lewis structure with eighteen electrons that also satisfies the octet rule is to have one double bond, one single bond, and one nonbonding pair of electrons. Using VSEPR, we conclude that the two oxygen atoms should occupy two of the vertices of a roughly equilateral triangle, with the nonbonding pair of electrons at the third vertex. This makes the bond angle equal to about 120°.

Polar and Nonpolar Molecules

If a molecule contains atoms of only one element, then the bonding is nonpolar covalent and the molecule as a whole is nonpolar. Examples include H_2, N_2, O_2, O_3, Cl_2, and P_4. Whenever a chemical bond is between atoms of different elements, the bond itself is polar covalent. Whether or not the entire molecule is polar, however, depends upon the symmetry of the molecule.

Symmetrical tetrahedral, triangular, and linear molecules are nonpolar because the polarities of the individual bonds cancel each other out when they are added together. In our examples, CH_4 and CO_2 are nonpolar. Unsymmetrical molecules, however, are usually polar—in our case, NH_3, H_2O, and SO_2. CH_3Cl is tetrahedral but polar because the polarities of the individual bonds do not cancel. Similarly, SO_3 is nonpolar, but S_2O_2, which has a sulfur atom that replaces one of the oxygen atoms, is polar. HCN is a linear molecule. Its Lewis structure of H–C≡N: shows C and N connected with a triple bond. Because the polarity of an H–C bond is less than the polarity of a C≡N bond, the polarities do not cancel, making HCN a polar molecule.

Because of the geometry of a pyramidal molecule, it is a safe conclusion that all pyramidal molecules are polar. Almost all bent molecules also are polar, with an exception being O_3, since it contains atoms of only one element.

Whether a substance is polar or nonpolar is one of the most important properties of that substance. It is an example of the old saying that oil and water don't mix. Oil molecules are nonpolar and water molecules are polar. Water, in fact, is extremely polar for such a small molecule, a property that accounts for the unusual properties water exhibits (see Chapter 24). To say that oil and water do not mix is to say that only like dissolves like. In other words, nonpolar substances tend to mix well together, polar substances tend to mix well together, but nonpolar and polar substances do not mix together well at all.

Intermolecular Forces

The forces that hold atoms together within a molecule (intramolecular forces) are fundamentally electrostatic in nature; the positively-charged nucleus of one atom is attracted to the negatively-charged electrons of another atom. Likewise, the forces of attraction between molecules (intermolecular forces) are also electrostatic in nature. This section is about three kinds of intermolecular forces: dipole-dipole forces, hydrogen bonding, and London forces.

Dipole-dipole Forces

There exists a separation of charge in polar molecules, not enough to cause ions to form, but enough to give one end of a polar molecule a slight positive charge and the opposite end a slight negative charge. Anytime positive and negative charges of the same magnitude are separated like that, a dipole (two poles—one positive and one negative) exists. On a time scale of pico- or femtoseconds, the distance between the poles fluctuates slightly; on a longer time scale the average distance between the poles is considered to be fixed.

When two polar molecules are in close proximity to each other—usually in liquids or solids—the negative end of one molecule feels an attraction to the positive end of the other molecule. Similarly, the positive end of the first molecule feels an attraction to the negative end of some other molecule, and the negative end of the second molecule feels an attraction to the positive end of some other molecule. This attraction is what we call a *dipole-dipole force* between molecules. In bulk liquids, the sum of all the dipole-dipole forces exerts a very strong force that is responsible for keeping the molecules together in the liquid phase, rather than the alternative of all the molecules evaporating.

Important Point

Different atoms in a molecule have different values of electronegativity. If the difference in electronegativity in a molecule is small, dipole-dipole forces are relatively weak. If a molecule contains one or more very electronegative elements——O, N, F, or Cl, for example——, dipole-dipole forces are much stronger.

Hydrogen Bonding

In some polar molecules, the presence of hydrogen atoms bonded to one or more atoms of very electronegative elements results in an enhanced intermolecular attraction called *hydrogen bonding*. When two molecules both contain an O–H, N–H, or F–H bond, there is hydrogen bonding between an oxygen, nitrogen, or fluorine atom on one molecule and a hydrogen atom on the other molecule. (Hydrogen atoms bonded to other elements do not exhibit hydrogen bonding.) Hydrogen bonding is a significantly stronger intermolecular force of attraction than dipole-dipole forces alone.

The most common molecules that exhibit hydrogen bonding are water (H_2O), ammonia (NH_3), and ethyl alcohol (C_2H_5OH). Organic molecules called *amines* are derived from ammonia, so they also exhibit hydrogen bonding. Since alcohols are defined as organic molecules that contain a C–O–H group, ethyl alcohol is only one of several alcohols that exhibit hydrogen bonding. Amino acids contain an amine group and a carboxyl group (COOH), so they exhibit hydrogen bonding. Water molecules can hydrogen bond to ammonia or alcohol molecules, which is why ammonia and alcohol are soluble in water.

When hydrogen bonding occurs between two molecules, the molecules both have to be polar. Therefore, in addition to hydrogen bonding, dipole-dipole forces are also present.

Historical Note

The term *hydrogen bond* is somewhat unfortunate, because a hydrogen bond is also the bond in an H_2 molecule. When hydrogen bonding was discovered in the 1930s, some chemists suggested using the term *hydrogen bridge* instead to distinguish it from the bond in an H_2 molecule. There are journal articles in the chemical literature in which the authors used the term *hydrogen bridge*. The term never really caught on, however, and has disappeared from the literature.

London Forces

Nonpolar molecules do not contain a dipole, at least a fixed dipole. However, there are forces of attraction between nonpolar molecules. Otherwise, nonpolar gases like CO_2, N_2, O_2, and He would never condense into liquids. The fact that at sufficiently cold temperatures a nonpolar gas does condense means there must be some kind of intermolecular force of attraction between its molecules. That force is too weak at high temperatures for the gas to condense, but at cold temperatures, it is strong enough to cause condensation. The same force is responsible for liquids freezing into solids at even colder temperatures.

This weak intermolecular force is caused by fluctuations in the electron distribution in a molecule that creates temporary, or instantaneous, dipoles. These instantaneous dipoles appear and disappear on a very short time scale. On a longer time scale, there is no fixed dipole and the molecule is nonpolar. However, these instantaneous dipoles exert just enough force on surrounding molecules to hold the molecules in the liquid or solid phase. This force goes by various names—London force, after the physicist Fritz London (1900–54), van der Waals force, after the scientist Johannes van der Waals (1837–1923), or dispersion force, because the electrical charges are dispersed through a molecule. The term *van der Waals forces* is also used by some scientists to refer to all varieties of intermolecular forces, so it will not be used here. My choice is to use the term *London force*, as that is the term used most often in chemistry.

Small nonpolar molecules have very weak London forces, so they exist most often as gases under conditions of room temperature and 1 atmosphere pressure. As the size of nonpolar molecules increases, London forces become stronger. For example, methane (CH_4) and octane (C_8H_{18}) are both nonpolar. Methane is a small molecule, has very weak London forces, and is a gas at room temperature. Octane is a larger molecule, has stronger London forces, and is a liquid at room temperature. As a general rule liquids with stronger London forces have higher boiling points than liquids with weak London forces.

The electrical charges in permanent dipoles fluctuate slightly in their distribution. When dipole-dipole forces are present between two molecules, those fluctuations also cause London forces to be present. In general the London forces will be weaker than the dipole-dipole forces, but both contribute to the attractions between molecules. If a molecule also exhibits hydrogen bonding, dipole-dipole and London forces will both be present and contribute to the attractions between molecules.

Resonance Structures

Chem Vocab

Resonance refers to the idea that the best representation of a molecule or polyatomic ion is a blend of Lewis structures in which multiple bonds are in different positions.

When drawing the Lewis structure for SO_2, it is an arbitrary decision to choose which oxygen atom gets the double bond to sulfur. The two oxygen atoms are equivalent and the double bond could go to either oxygen. Therefore, there are two possible Lewis structures for SO_2, both of which are equally valid. In fact, the double bond tends to move around, or resonate around, the molecule. Chemists call this phenomenon delocalization of electrons.

The best way to picture a molecule like SO_2 is as a structure that is a blend of the two resonance structures. A snapshot of the molecule taken on a timescale of a picosecond would show just one of the structures. But, a picture taken on a timescale of a microsecond would show a blur of electrons with equal electron density between the sulfur atom and each of the two oxygen atoms.

Expanded Octets

There are a large number of molecules in which the central atom bonds to more than just four other atoms. Obviously, that would mean more than eight valence electrons around the central atom, which violates the octet rule. In these cases we say that the central atoms are exhibiting expanded octets.

Important Point

Resonance, or delocalization of electrons, takes place because it gives a molecule or ion the structure with the lowest possible energy.

Among compounds containing nonmetals, the most common structures are ones with five or six atoms attached to the central atom. Typical shapes of molecules in these cases are trigonal bipyramids in the case of five atoms, and octahedra in the case of six atoms. Since these molecules less commonly occur naturally, their existence will not be discussed further. In this book, most of our examples of compounds containing nonmetals have carbon, nitrogen, oxygen, or sulfur as the central atoms, in which case we will always assume that the octet rule holds.

Hybrid Orbitals

Our simple pictures in Chapter 16 of s, p, d, and f orbitals are not satisfactory to explain the bond angles in most molecules—the angles of 109°, 120°, and 180° we have found in most of our examples in this chapter. The orbitals shown in Chapter 16, however, are for single isolated atoms. When atoms come together to form molecules, their atomic orbitals blend together. We call these orbitals hybrids of the atomic orbitals.

Important Point

Hybrid orbitals form because they represent the lowest-energy arrangements of atoms possible.

For example, to explain the tetrahedral geometry of CH_4, we say that carbon's 2s and three 2p orbitals blend together, or hybridize, to form four sp^3 hybrid orbitals. (Four atomic orbitals combine to make four hybrid orbitals.) The four sp^3 orbitals, following VSEPR, orient themselves toward the vertices of a regular tetrahedron, hence methane's tetrahedral shape.

With SO_2, sulfur's 3s orbital and two of its 3p orbitals blend together to form three sp^2 orbitals that orient themselves toward the vertices of an equilateral triangle. Two of the sp^2

hybrids overlap with the orbitals on the oxygen atoms and one sp^2 hybrid is occupied by the nonbonding pair of electrons.

With CO_2, carbon's 2s orbital and just one of its 2p orbitals combine to make two sp hybrids that move 180° apart.

Atoms tend to come together so as to result in molecules with the lowest possible energy. Whether a central atom forms sp, sp^2, or sp^3 hybrids depends upon the kinds of elements that are in the molecule and how many atoms there are.

Shapes of Larger Molecules

With larger molecules than the ones we have been considering, we usually do not specify a shape for the entire molecule. Instead, we describe the shape around each of the main atoms in the structure. For example, in methyl alcohol we say that the shape around the carbon atom is tetrahedral and the shape around the oxygen atom is bent.

Conclusion

Alchemists and eighteenth and nineteenth-century chemists were unable to explain why various substances have the properties that they do. It was only in the twentieth century that chemists began to describe the shapes of molecules and to relate macroscopic properties to microscopic shapes. That understanding has resulted in the ability of modern chemists to design novel molecules with various desired properties. Once again, better living through chemistry!

CHAPTER 21

The Properties of Gases

In This Chapter

➤ Units of measurement of gases

➤ Gas laws

➤ The ideal gas law

➤ The Kinetic Theory of Gases

The air we breathe is a mixture of many gases. Mostly air consists of N_2, O_2, Ar, and water vapor. However, air also contains CO_2 and numerous trace gases, including NO, NO_2, SO_2, SO_3, NH_3, CH_4, H_2S, He, and many more. This chapter is about the properties of gases.

Units of Measurement

Table 21.1 shows the common units of measurements used in the study of gases.

Table 21.1: Units of Measurement of Gases

Variable	Common Units	Conversions
Pressure (P)	atm, mm Hg, torr, in Hg, bar, mb, lb/sq in	1 atm = 760 mm Hg 1 mm Hg = 1 torr
Volume (V)	m^3, cm^3, Liter (L), mL	1 L = 1000 mL 1 mL = 1 cm^3 1 m^3 = 1000 L
Temperature (T)	°F, °C, K	$T(°F) = 1.8T(°C) + 32°$ $T(K) = T(°C) + 273$
Quantity (n)	mole	n/a

During the seventeenth, eighteenth, and nineteenth centuries, scientists performed a number of experiments to determine the relationships among the variables listed in Table 21.1. The laws of gases are usually referred to by the names of their discoverers and are called Boyle's law, Charles's law, Avogadro's law, Dalton's law, Graham's law, and the ideal gas law.

Boyle's Law

Robert Boyle (1627–91) was an Irish natural philosopher who studied the quantitative relationship between the pressure and volume of a gas. What he discovered was that the product of pressure times volume (P x V) is a constant if the quantity of gas (n) and the temperature (T) are both held constant.

Oftentimes, Boyle's law is also stated as $P_1V_1 = P_2V_2$. This says that when the quantity of gas and the temperature of the gas remain fixed, the volume of a gas is inversely proportional to the pressure applied to the gas. Increase the pressure (squeeze hard on a balloon) and the volume decreases (the balloon becomes smaller); decrease the pressure and the volume increases.

Important Point

Boyle's law states that pressure times volume is a constant for a fixed quantity of gas. In other words, as pressure increases, volume decreases; as pressure decreases, volume increases. This is only true, however, if the temperature of the gas stays constant. Boyle's law only applies to gases in balloons or in other containers in which volume can change, not in containers with rigid walls like a metal box.

Charles's Law

Jacques Alexandre César Charles (1746–1823) was a French scientist who was interested in both the behavior of gases and the behavior of balloons, the latter of which were just being developed when Charles was in his early thirties.

In his study of gases, Charles discovered that the volume of a gas is proportional to its temperature if the quantity of gas (n) and the pressure (P) are held constant. Quantitatively, we can write

$$V = kT, \text{ or } \frac{V_1}{T_1} = \frac{V_2}{T_2}.$$

This says that when the quantity of gas and the pressure on the gas remain fixed, the ratio of a gas's volume to the gas's temperature is a constant. Consequently, if you heat a gas (immerse a balloon in boiling water), the volume of the gas increases (the balloon inflates); decrease the temperature (immerse the balloon in ice water) and the volume decreases (the balloon deflates).

Important Point

Charles's law states that the ratio of volume to temperature is a constant for a fixed quantity of gas. In other words, as a gas gets hotter, its volume expands; as a gas cools, its volume contracts. This is only true, however, when the pressure and quantity of the gas stay constant.

The Need for an Absolute Temperature Scale

At first glance it would appear that when the temperature of a gas is increased from 20°C to 40°C (or any other doubling on the Celsius scale), the volume would double, assuming pressure stays constant. Experiments show, however, that the volume only increases by a factor of 1.07, not 2. What is going on here?

The answer is that most measurement scales with which we are familiar are absolute scales in the sense that 0 on a scale means the absence of whatever it is we are measuring. Celsius (or Fahrenheit) temperatures, however, are not absolute temperatures. The temperature of an object is a measure of the object's heat content. A temperature of 0 degrees on either scale does not indicate the absence of heat. Therefore, multiplying or dividing by a Celsius or Fahrenheit temperature is meaningless in any real physical sense. Instead, we must convert temperature to an absolute scale. Kelvin temperatures

Important Point

There is a second absolute temperature scale that most often is used in engineering. Just as the Kelvin scale is the absolute temperature scale based on the Celsius scale, the Rankine temperature scale is the absolute temperature scale based on the Fahrenheit scale. On the Rankine scale, 0 = –460°F.

are absolute values based on Celsius temperatures. So, before we multiply or divide by temperature, we must convert degrees Celsius to Kelvins by adding 273° to the Celsius temperature. In the present example, 20°C = 293 K and 40°C = 313 K. The ratio of 313 to 293 is 1.07.

Chem Vocab

For many years the Celsius scale was called the centigrade scale. Recently, the decision was made to name all temperature scales after their creators, although you may sometimes still see centigrade in use. It is also conventional to use the degree (°) symbol with Celsius and Fahrenheit temperatures, but not with Kelvin temperatures, although you will probably see °K used by some authors.

Avogadro's Law

Amedeo Avogadro (1776–1856) was an Italian physicist. His famous hypothesis about gases was that "at the same pressures and temperatures, equal volumes of gases contain equal numbers of particles." Chemists and physicists in Avogadro's time were still undecided about the existence of atoms and molecules. Without experimental evidence to support his hypothesis, Avogadro was pretty much ignored during his lifetime.

Eventually, Avogadro was shown to be correct, and today we call his hypothesis a law and write it in the form $V = kn$ when P and T are held constant. We also write Avogadro's law in the form $\frac{V_1}{n_1} = \frac{V_2}{n_2}$. To say, however, that equal volumes of gases at the same pressure and temperature contain an equal number of gas particles does not tell us how many gas particles there are. Certainly in the nineteenth century nobody knew of any way to count atoms

Historical Note

Avogadro died many years before the value of Avogadro's number was discovered. That determination of Avogadro's number had to wait until the early twentieth century when California physicist Robert Millikan measured the charge on the electron. Since scientists knew the charge on a mole of electrons, the ratio of the two numbers yielded Avogadro's number—6.02×10^{23}.

or molecules, assuming one even believed in them. Eventually, though, the question of trying to count the number of particles in a sample of gas led to the concept of Avogadro's number—the number of particles in a mole of particles.

Historical Note

Avogadro died many years before the value of Avogadro's number was discovered. That determination of Avogadro's number had to wait until the early twentieth century when California physicist Robert Millikan measured the charge on the electron. Since scientists knew the charge on a mole of electrons, the ratio of the two numbers yielded Avogadro's number—6.02×10^{23}.

The Ideal Gas Law

In the mid-nineteenth century, scientists combined Boyle's law, Charles's law, and Avogadro's law into a single law called the ideal gas law. Usually, the ideal gas law is stated in the following form:

$$PV = nRT$$

where R is the gas constant. The numerical value of R depends upon the units used for P, V, n, and T. A common set of units in chemistry is to measure P in atmospheres, V in liters, n in moles, and T in kelvins. Then R is equal to 0.0821 atm L/mol K.

Historical Note

In the 1800s, the ideal gas law was deduced empirically by combining Boyle's law, Charles's Law, and Avogadro's law into a single equation. However, the ideal gas law was also derived mathematically from Newton's laws of motion. The agreement of the two methods was a striking success for mid-nineteenth-century physicists.

Definition of an Ideal Gas

Let us rearrange the ideal gas law into the following form:

$$P = \frac{nRT}{V}$$

Now suppose we set up a container of known volume V with a known quantity of gas n. We measure the temperature using a thermometer and the pressure using a pressure gauge. Now we calculate the pressure using the above equation. If the calculated pressure and the pressure measured by the gauge agree, then we say the gas is exhibiting ideal behavior. What makes the ideal gas law so useful is that most gases—especially common ones like N_2, O_2, H_2, CO_2, He, and the other noble gases— exhibit ideal behavior over a wide range of pressures and temperatures. Under exceptional conditions—extremely high pressures or very low temperatures not often encountered—gases deviate from ideal behavior. We can ignore those exceptional conditions for our purposes.

Important Point

Since air is 78 percent N_2, 21 percent O_2, and 1 percent Ar, the average molecular weight of air is about 29 g/mol. According to the principles of buoyancy, any gas with a molecular weight that is less than 29 g/mol will have lifting power compared to air. For all practical purposes, that limits us to H_2 (2.02 g/mol) and He (4.00 g/mol). Since H_2 is explosive, that really only leaves us with He with which to fill balloons (and blimps and dirigibles). Any other gas is too heavy to exhibit any buoyancy.

The Kinetic Theory of Gases

The wide utility of the ideal gas law suggests that the gaseous state is a relatively simple state of matter to understand and that the physical properties of all ideal gases tend to be pretty much the same. Chemists and physicists summarize the physical properties of gases in a model that was developed in the middle of the nineteenth century, the kinetic theory of gases.

The postulates of the kinetic theory are as follows:

➤ A gas consists of small particles moving in continuous, random, chaotic motion.

➤ Gas particles are so tiny that if they were all packed together into a little ball, the volume they would occupy would be completely negligible compared to the volume of the container they are in.

➤ Gas particles are constantly colliding with each other and with the walls of their container. These collisions are similar in behavior to balls on a billiard table. In any collision of two objects, total energy and total momentum are always conserved, but in this case kinetic energy is also conserved. Physicists call this type of collision an elastic collision.

➤ The gas particles move with varying kinetic energy. The number of particles possessing each possible kinetic energy can be graphed. The graph shows a statistical distribution of energy called the Maxwell-Boltzmann distribution after the British physicist James Clerk Maxwell (1831–79) and the Austrian physicist Ludwig Boltzmann (1844–1906), who developed the distribution independent of each other.

The shape of the graph is independent of the gas that is being graphed. At the same temperature all gases have the same distribution of kinetic energy. This means that if two gases are at the same pressure, they have identical average kinetic energies. The average kinetic energy of a gas $\overline{K}$ is given by the following equation:

$$\overline{K} = \frac{3}{2} R\, T, \text{where} R = 8.31 \frac{J}{mol \cdot K}$$

The kinetic theory of gases gives us another way of defining an ideal gas: A gas is said to be ideal if its properties satisfy the postulates of the kinetic theory of gases.

Dalton's Law of Partial Pressures

We have encountered the early nineteenth-century English chemist John Dalton before in the discussion to his development of the theory of atoms. Dalton also worked on the properties of gases.

In a mixture of gases, the partial pressure of any one gas is the pressure that gas alone would exert if the other gases were removed (and volume and temperature were kept constant). Chemists had already noticed that the volume of liquids is not additive. For example, if 100 mL of one liquid is mixed with 100 mL of another liquid, the total volume is not exactly 200 mL—the total usually is a little less or a little more than 200 mL. Therefore, it was not clear that the pressure of two gases in a mixture should be additive. It was Dalton who argued that they are.

Dalton's law of partial pressures states that "The total pressure of a mixture of gases equals the sum of the partial pressures." It may sound simple to us today, but remember that Dalton lived before the ideal gas law or the kinetic theory of gases had been formulated.

Chem Vocab

The partial pressure of a gas is the pressure that one component of a mixture of gases would continue to exert if all of the other gases were removed, keeping volume and pressure constant.

How We Calculate Partial Pressure

A pressure gauge only indicates total pressure, not the partial pressure due to each gas in a mixture. However, we can use the ideal gas law in the form $P = \frac{nRT}{V}$ to calculate each partial pressure. If only one gas is present, n equals the number of moles of that gas (found by dividing the mass of the gas by the molecular weight of the gas). By the kinetic theory of gases, the pressure a gas exerts is independent of the identity of the gas. All that matters is the number of particles of that gas that are present. In a mixture of gases, we know the volume and temperature. Therefore, if we know the mass of the various gases that have been mixed together, we can likewise calculate the number of moles of each gas present. Then, the partial pressure of each gas is just the number of moles of that gas times $\frac{RT}{V}$.

Note that when we add all the partial pressures together, all of the n's will add together to give the total number of moles of gases present. Thus, the sum of all of the individual partial pressures equals the total gas pressure that would be measured on a pressure gauge—Dalton's law of partial pressures.

Another way to state this is to use the concept of mole fraction. In a mixture of two substances A and B, the mole fraction of A = $\frac{\text{\# moles of A}}{\text{total \# moles}}$, and the mole fraction of B = $\frac{\text{\# moles of B}}{\text{total \# moles}}$. (Note that the sum of the mole fractions must always equal 1.00.) Then the partial pressure of gas A is just the mole fraction of A times the total pressure.

Chem Vocab

Diffusion is the spreading out of a substance into a container so that it mixes uniformly with the other substances in the container.

Effusion is the passage of a gas through a pinhole into a vacuum.

Graham's Law of Effusion

First, some definitions. You have probably heard of diffusion, which is the spreading out of a substance (perfume, for example) into a container. Effusion is a less familiar term. Effusion refers to the situation of dividing a container into two compartments separated by a wall. A pinhole is made in the wall. One compartment is filled with a gas. The other compartment is evacuated. Remember that gas particles are moving in continuous, random, chaotic motion. Once in a while, a gas particle passes through the pinhole into the evacuated side. The pinhole is so tiny that only one molecule can pass through at a time.

Thomas Graham (1805–69) was a Scottish chemist. In studying the effusion of gases, Graham experimentally compared the time required for two samples of different gases containing the same number of particles to effuse from one side of a container to the other. Keeping temperature constant, he found that the time for each gas was directly proportional to the square root of its molecular weight (MW):

$$\text{Time} \propto \sqrt{\text{MW}},$$

where $\propto$ is the symbol that means "is proportional to." If we write the ratio of the time of effusion for two gases (label them A and B), we can replace the proportionality symbol with an equal sign, as shown:

$$\frac{T_A}{T_B} = \sqrt{\frac{MW_A}{MW_B}}.$$

Equivalently, since the time required for a gas to effuse is inversely related to the average speed, or rate, at which the gas is traveling, we can also write an expression for the relative rate (R) of effusion of two gases that are at the same temperature:

$$\frac{R_A}{R_B} = \sqrt{\frac{MW_B}{MW_A}}.$$

Conclusion

Nineteenth-century scientists made remarkable discoveries about gases considering that in most cases they could not even see the gases with which they were working. The laws governing the properties and behavior of gases have served us well. It may seem peculiar to learn about laws that were first stated over one hundred years ago, but the gas laws are firmly established as part of the foundation of the physical sciences.

CHAPTER 22

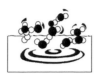 **Solutions and How We Describe Them**

In This Chapter

➤ Aquatic environments as solutions

➤ The atmosphere as a solution

➤ Examples of solutions of solids, liquids, and gases

A solution is a homogeneous mixture that contains two or more elements or compounds mixed together in uniform proportions. The phrase *uniform proportions* means that if you extract samples of a solution, you will get the same relative amount of the components regardless of where the sample was drawn. For example, if you have a gallon bottle of sugar water that you have prepared to fill your hummingbird feeders, a cup of the solution drawn from the bottom of the bottle will have the same concentration of sugar as a cup skimmed from the surface.

Solutions: Solutes and Solvents

A solution is a mixture containing two or more elements or compounds. Whichever element or compound is present in greatest quantity is called the solvent. A solution could have several solutes, or other components. Bottled beverages are good examples. Water is the solvent, but read the label and you will see that there are several solutes that include flavorings and preservatives. In the discussions that follow, the examples may have only a single solute. In nature, however, it is more common to have several solutes. The two most common solutions in nature are bodies of water—lakes, ponds, rivers, and oceans—and the atmosphere.

Chem Vocab

The solvent is the component of a solution that is present in greatest amount.

The solutes are all of the other components of a solution.

Chem Vocab

Anthropogenic means "caused by humans." *Anthropo* means "human" and *genic* means "produced by."

Important Point

The atmosphere is a mixture of gases, the most abundant of which is nitrogen, which makes N_2 the solvent. The other gases—O_2, Ar, CO_2, water vapor, and trace gases—are the solutes.

Solutions: Aquatic Environments

A large number of substances may be dissolved in aquatic bodies, some natural, some anthropogenic. To describe an ocean as being full of salt water is not to suggest that there is only one salt (table salt, or NaCl, for example) in ocean water. There are many salts, as well as "non-salt" compounds like dissolved carbon dioxide. Salts are brought to the oceans by rivers, which in turn drain lakes and ponds. The concentration of salts and other dissolved substances in freshwater environments are usually much less than they are in the oceans, but they are there nevertheless.

Solutions: The Atmosphere

The atmosphere is a huge mixture of gases. Since the relative concentration of gases in dry, unpolluted air is constant, the atmosphere may be considered to be a solution. (The number of gas molecules decreases with increasing altitude, but the relative number of the different gases stays the same.) The reference to a "dry" atmosphere is because this discussion will neglect water vapor. Water vapor in the atmosphere is too highly variable and is usually not included in lists of the gases found in the atmosphere. The atmospheric gas present in greatest amount is nitrogen (N_2), so that is the solvent. All of the other gases—O_2, Ar, CO_2, etc.—are the solutes.

Other Kinds of Solutions

Solvents can be solids, liquids, or gases; solutes can be solids, liquids, or gases. The example of aquatic environments illustrates the kind of solution that is most familiar to people—solids (and gases) dissolved in a liquid. Water is our most common liquid, so it is our most common solvent. The example of the atmosphere illustrates a solution in which several gases are dissolved in another gas. There are other examples of solutions.

In addition to solids and gases dissolving in liquids, there are solutions in which liquids are dissolved in other liquids. When two liquids are mutually soluble in each other in all proportions, we say the two liquids are miscible. (If they are not soluble in each other, we say they are immiscible.)

Chem Vocab

Miscible means that two liquids dissolve into each other in all proportions.

Alcoholic beverages are a familiar example. Normally we think of water as the solvent and ethyl alcohol as the solute. The proof of an alcoholic beverage is two times the concentration of alcohol. Thus, a 40-proof beverage is 20 percent alcohol and 80 percent water. A 100-proof beverage is 50 percent alcohol and 50 percent water. If the proof were greater than 100, that would mean there is more alcohol than water, so their roles as solute and solvent would be reversed—ethyl alcohol would become the solvent and water would be the solute. Many medications dispensed in liquid form are also examples of liquids dissolved in another liquid.

In addition to gases dissolving in another gas, as is the case with the atmosphere, solids and liquids can also dissolve in gases. An example of solids that dissolve in gases is again the atmosphere, which frequently has dust, soot, or other fine particulate matter suspended in it. Winds serve to maintain the suspension, but in absolutely still air, solid materials settle out due to the force of gravity.

Water droplets can also be suspended in the atmosphere, which is an example of a liquid dissolved in gases. When the water droplets become large enough, it rains and gravity causes the droplets to fall to Earth.

Less obvious are examples in which a solid serves as the solvent. Alloys consist of two elements (often two metals) that have been mixed together uniformly. Whichever element is present at a greater amount is the solvent.

As water freezes, ice can trap liquid or gas bubbles in it. Lava can do the same thing. As the lava cools and solidifies, various solids, liquids, or gases can be trapped in it. Pumice is a good example of volcanic material that cooled so rapidly that dissolved atmospheric gases did not have time to escape, so the gases were trapped inside. As a result, pumice has a very low density (it floats on water) and very fine grains.

Chem Vocab

An alloy is a homogeneous mixture of two or more metals or elements that have metallic properties.

On the other hand, volcanic material may cool down so slowly that all of the gases do have time to escape. The result is the formation of rocks like basalt that are very hard and dense compared to pumice. Of course, there are several rocky materials that can form in-between these two extremes and that can have varying mineral compositions and grain sizes, resulting in various other igneous rocks such as granite, diorite, gabbro, rhyolite, and obsidian.

Saturated Solutions

At a given temperature and pressure, any solvent can hold at most a certain maximum amount of solute. If the solute is at its maximum concentration, the solution is said to be saturated. If the solute is present at a concentration that is less than the maximum possible, the solution is said to be unsaturated. In both situations, the solution is in a state of equilibrium. If temperature and pressure remain constant, the solution can be at equilibrium indefinitely.

Chem Vocab

Saturated means "unable to hold or contain any more," and in the case of solutions it means "unable to contain any more solute under the prevailing conditions of temperature and pressure."

A supersaturated solution is one that is temporarily more concentrated than should be possible for a given combination of temperature and pressure.

Sometimes solutions can temporarily hold more than the maximum quantity of solute possible at that temperature and pressure. In that case the solution is said to be supersaturated. This is not a stable equilibrium situation. Eventually, solid will precipitate until the solution is saturated.

Supersaturation sometimes occurs when ionic solids are dissolved in water. The concentration of the ions may be more than the solution should be able to hold, but precipitation has not yet occurred. Ions usually need little specks of crystals or even dust particles to cling to in order to precipitate. Dropping in a tiny crystal or a speck of dust or even shaking the container usually results in precipitation.

The same situation may exist in a very clean atmosphere. Rain droplets tend to form on dust particles in the atmosphere. If there is no dust, there could be enough water vapor in the air that it should rain, but it does not. Cloud seeding works this way. By adding salt crystals to the atmosphere, nucleation is facilitated, rain droplets form, and it rains. The salt crystals need to be of a size conducive to water droplet formation—silver iodide (AgI) is commonly used.

Chem Vocab

Nucleation refers to the formation of tiny bits of a solute around which more solute can precipitate.

Conclusion

Solutions are very much a part of our daily lives. A bottle of almost any liquid sold at a supermarket, pharmacy, hardware store, or garden center is a solution, usually with several components. Reading labels can save you money! Compare the ingredients listed on the labels and buy the products with the ingredients you want and are the least expensive.

CHAPTER 23

Physical Properties of Solutions

In This Chapter

➤ How polarity affects solubility

➤ How temperature and pressure affect solubility

➤ How a solute affects a liquid's vapor pressure

➤ How a solute affects a liquid's freezing and boiling point

Many of a solution's properties are determined by the molecular structures of the substances that are in them. In this chapter we will explain the principles underlying a number of familiar everyday phenomena.

The Effect of Polarity on Solubility

As was emphasized in Chapter 20, whether a molecule is polar or nonpolar is one of the most important properties of a molecule. You have probably heard the saying: "Oil and water don't mix." That is because oil is nonpolar and water is polar. Chemists have a saying that like dissolves like. That is, polar substances tend to be soluble in other polar

---

Important Point

Like dissolves like. Polar substances tend to be soluble in other polar substances; nonpolar substances tend to be soluble in other nonpolar substances. Polar and nonpolar substances, however, tend not to mix.

---

substances, nonpolar substances tend to be soluble in other nonpolar substances, but polar and nonpolar substances do not mix. Let us examine why that is true, using oil and water as an example.

Mixing Polar and Nonpolar Substances

Oil molecules are attracted to each other by London forces. Because oil molecules are fairly large, the London forces are fairly strong. Water molecules, on the other hand, are attracted to other water molecules by hydrogen bonding. Hydrogen bonding is an extremely strong force, so the attraction between water molecules is very strong.

When we try to mix oil and water, the oil molecules have a much stronger attraction to other oil molecules than they do to water molecules, so the oil molecules tend to stick together. Likewise, water molecules have a much stronger attraction to other water molecules than they do to oil molecules, so water molecules tend to be surrounded by other water molecules. We might say that the water molecules push out the oil molecules.

Because oil is less dense than water, the oil will float on top of the water. The boundary between the two is quite distinct. Oil and water don't mix. They are immiscible. No amount of mixing will change the immiscibility of oil and water. They really do not form a solution. It is a mixture and the two components are clearly distinct.

Mixing Two Polar Substances

In contrast to oil, several alcohols (ones with small molecules) are completely miscible with water. Alcohol molecules are polar, and they exhibit hydrogen bonding. Because water also exhibits hydrogen bonding, the force of attraction between alcohol molecules, the force of attraction between water molecules, *and* the force attraction between alcohol and water are all of comparable strength. Therefore, alcohol molecules are just as likely to be surrounded by water molecules as by other alcohol molecules. The two liquids mix completely and there are no visible boundaries between them. Small alcohols (containing one to four carbon atoms) form true solutions with water.

Chem Vocab

An ion that is hydrated is surrounded by water molecules rather than by other ions.

Salts consist of positive and negative ions and are polar. Therefore, if a salt is going to be soluble in any liquid, it is much more likely to be soluble in water than in nonpolar organic liquids. All sodium salts, for example, are soluble in water. In addition to the polarity, sodium ions (Na^+) themselves have a strong affinity for water molecules. When a sodium salt dissolves in water, the two ions separate and the Na^+ ion becomes completely surrounded by water molecules. (We say that the Na^+ ion is hydrated.) Likewise, any salt containing K^+ and the other alkali metals NO_3^- or NH_4^+ is also soluble in water. The attraction between the two ions in the salt is less than the attraction the ions have for water molecules, so the salt dissolves. When a salt is insoluble in water, it means that the attraction between the two ions is greater than the attraction between the ions and water, so the salt remains in solid form.

Mixing Two Nonpolar Substances

Most organic solvents are nonpolar, so they are immiscible with water. However, most organic solvents are miscible with each other. An example of a nonpolar organic solvent is carbon tetrachloride (CCl_4). We said that oil and water don't mix. Oil and carbon tetrachloride, however, do mix—they are completely miscible. London forces are at work here.

The relative strength of the London forces between oil molecules, between CCl_4 molecules, and between oil and CCl_4 are all of comparable strength. Therefore, oil molecules are just as likely to be surrounded by CCl_4 molecules as they are to be surrounded by other oil molecules, and vice versa. Oil and CCl_4 form a true solution.

Examples of the Effect of Polarity on Solubility

The purpose of soap (and detergent) is to overcome the lack of attraction between oil and water molecules. A soap molecule is a very long molecule. One end of a soap molecule is ionic and therefore polar. Because it mixes with water, the polar end is said to be hydrophilic (water loving). The other end is nonpolar. Because the nonpolar end does not mix with water, it is said to be hydrophobic (water avoiding). When you wash your hands, the polar end of the soap molecule dissolves in water and the nonpolar end dissolves in the oil, bringing the oil and water together and leaving your hands clean.

Chem Vocab

Hydrophilic means "water loving," or "tends to mix with water." Small sugar molecules and many salts are hydrophilic.

Hydrophobic means "water avoiding," or "does not mix with water." Fats and oils tend to be hydrophobic.

Different kinds of paint also demonstrate the effect of polarity on solubility. Oil-based (enamel) paints are nonpolar. You would not want to try to clean your paint brush afterward with soap and water. It would make a mess. Paint stores sell paint thinner, which is a mixture of nonpolar organic solvents that do dissolve oil-based paints. Other nonpolar solvents that may be used are acetone or turpentine. Latex paints, on the other hand, are water based and therefore polar, so they easily clean up with soap and water. Historically all paints were oil based. In Mark Twain's story of Tom Sawyer and the white-washed fence, the white paint would have been oil based. Latex paints were developed to make cleanup easier.

Fingernail polish remover contains either acetone or ethyl acetate, which are both nonpolar. Acetone is a little harder on the skin, so some people prefer ethyl acetate. Fingernail polish remover has other uses around the house. Whenever I try to remove the price tag from a book, for example, and it leaves a gummy residue, I use fingernail polish remover to eliminate the remaining residue. (Ethyl acetate seems to work just a little bit better than acetone does for this purpose.) Of course, I apply only a small amount first, just to make sure the ethyl acetate is not going to discolor the book's cover. I do not recall that it ever has, but it does not hurt to check first. Fingernail polish remover can also be used to remove ink or other stains from smooth surfaces if other cleaners do not work.

Important Point

Before using any chemicals around the house, the garage, or the yard, read the instructions. Many commercial products should be used only when wearing gloves or a face mask. The fumes of some are noxious enough they should only be used in well-ventilated spaces. Some members of your household could have specific allergies or sensitivities to chemicals. Most cleansers should be kept away from children and pets. Many products could damage painted surfaces or fabrics and should be tested on an inconspicuous piece of material before full treatment.

Other household cleaners are water based and therefore polar. Window cleaners are a good example. Window cleaners often contain ammonia, which is a weak base. Vinegar, which contains the weak acid acetic acid, works well, too. Acids and bases tend to be polar, which is why they are soluble in water.

Another nonpolar solvent I have used at home is xylene, which is actually a mixture of xylene isomers. Xylene would probably damage indoor surfaces, so I recommend only using it outdoors. It works well, for example, removing heavy grease stains from concrete.

The Effect of Temperature on Solubility

Recall that all chemical reactions are either endothermic or exothermic; when the reaction takes place, heat must either be absorbed from the surroundings or given off to the surroundings. The same is true when solutes go into solution. The solution process generally is either endothermic or exothermic. An exception is the solution of gases in other gases. Because there is no real interaction between the gas molecules, that process is neither endothermic nor exothermic.

Important Point

If a solute needs to absorb heat in order to dissolve, it will be more soluble at higher temperatures. If a solute needs to give off heat in order to dissolve, it will be more soluble at lower temperatures.

Dissolving Solids in Liquids

When a solid dissolves in a liquid, the process could be either endothermic or exothermic. Consider what happens when a soluble salt dissolves in water. In most cases heat must be absorbed from the surroundings. Table salt, for example, absorbs heat in order to dissolve in water. We can show this using the following equation:

$$NaCl \ (s) + water + heat \rightarrow NaCl \ (aq),$$

where the symbol *aq* reminds us that we now have an aqueous solution of NaCl. Because this solution process is endothermic, it makes sense that raising the temperature should increase the solubility of NaCl in water, and it does. In this case we would say that as the temperature increases, the solubility also increases.

Dissolving sugar in water is another example of an endothermic process. Hot coffee drinkers are probably aware of the fact that the solubility of sugar increases at higher temperatures.

Important Point

It is more common for the solubility of solids in water to increase with increasing temperatures than for solubility to decrease with increasing temperatures.

If raising the temperature increases solubility, then decreasing the temperature should decrease solubility. This is what happens when lava cools. Many of the minerals that deposit from lava beds do so because they become less soluble at cooler temperatures. Crystals begin to grow and will do so until the supply of ions needed is finally depleted.

On the other hand, consider what happens when $CaCO_3$ dissolves in water. This time the process is exothermic, which we can show in the following equation:

$$CaCO_3 \text{ (s)} + water \rightarrow CaCO_3 \text{ } (aq) + heat.$$

Since the system must give off heat in order for $CaCO_3$ to dissolve, increasing the temperature works against the solution process. This time, as the temperature increases, the solubility decreases; $CaCO_3$ becomes less soluble. In solution processes that are exothermic, lowering the temperature—not raising the temperature— increases solubility.

Important Point

The white scale that builds up inside coffee pots, tea kettles, hot water heaters, and automobile radiators, or around an automobile's battery terminals is calcium carbonate. $CaCO_3$ is a base and is soluble in acid. I have seen people clean their car or truck's battery terminals by pouring a bottle of coke over them. Soaking the inside of a tea kettle or coffee pot overnight with a vinegar solution should remove most of the scale. Auto parts stores sell products for removing scale from radiators without harming the radiator itself.

You are probably familiar with the insolubility of $CaCO_3$ at high temperatures. Calcium (Ca^{2+}) and carbonate (CO_3^{2-}) or bicarbonate (HCO_3^-) ions are commonly found in tap water. Since water in tea kettles and coffee pots are heated to a high temperature, $CaCO_3$ precipitates as a white scale inside the kettles or pots. Since carbonate-containing compounds dissolve in acid, soaking the inside of kettles or pots with vinegar should dissolve the scale. Similarly, scale can build up inside water heaters. Too much scale begins to insulate the inside walls of the heaters, weakening the walls and eventually causing them to leak. When a water heater has to be replaced, the culprit is likely to be the buildup of $CaCO_3$. Another example is your car's radiator. Because the water flowing into the radiator has been heated by the engine, $CaCO_3$ precipitates inside radiators also.

Dissolving Gases in Liquids

To the best of my knowledge, the process of a gas dissolving in a liquid is always exothermic. Therefore, all gases are less soluble at a higher temperature. Most gases are not very soluble in water anyway, but they are even less so when heat is added. In fact, dissolved gases can be driven out of water entirely by boiling the water. Here are just a few examples of the way in which higher temperatures lower the solubility of dissolved gases.

Perhaps the most familiar example is carbonated beverages. Soft drinks are canned or bottled under pressure so that they contain dissolved CO_2. Cans or bottles quickly lose their fizz when opened because the pressure decreases. At room temperature, however, they lose their fizz even more rapidly because CO_2 is less soluble at room temperature than at the temperature of the inside of a refrigerator.

Another example of the lower solubility of gases at higher temperatures is the effect of temperature on the solubility of oxygen gas in water. Aquatic organisms obtain oxygen through their gills by assimilating oxygen that is dissolved in the body of water in which they live. Warm water is not a conducive environment for many organisms. Trout, for example, prefer cold-water environments because cold water holds more oxygen than warm water does. In temperate climates, as temperatures begin to drop during autumn, lakes and ponds become colder and may eventually freeze. As the temperature drops, the concentration of oxygen gas in a lake or pond increases. If the lake freezes over, the supply of oxygen present before freezing occurred is all the fish, or any other aquatic organism, have to live on until the ice thaws in the spring. One of the concerns biologists have about global warming is the decrease in dissolved oxygen in aquatic environments as those environments become warmer. Although some aquatic species may prefer warmer waters, other species will suffer.

Important Point

Clear-cutting of forests along streams has always been a problem for fish and other aquatic organisms that live in those streams. The removal of trees increases erosion, causing sediment to wash into the streams and possibly clogging the gills of fish, amphibians, and aquatic invertebrates. And without the shade that the forest canopy was providing, the temperature of the water may increase to the point that it no longer contains enough dissolved oxygen for aquatic organisms to survive.

The decreased solubility of O_2 at higher temperatures can have deleterious effects on fish species that migrate up rivers. Fossil-fueled or nuclear electrical power plants may be located along those rivers. These power plants run their generators by burning coal, oil, or natural gas, or by the fission of uranium. The heat that is produced converts water into steam, which spins the turbines, which in turn spin the generators. Power plants usually are built next to bodies of water to utilize water in the condensing portion of the cycle. The heated water is returned to the river where it continues to flow downstream.

For a short distance downstream from the power plant, the water is unnaturally warm, which means it may be depleted of dissolved O_2. Fish that are trying to migrate upstream to spawn may be unable to swim through such warm water. A warm section of river can be as much of a barrier to fish migration as a dam is.

The solution is to build a channel for the fish that bypasses the power plant. It is a solution similar to the one often used when a dam is blocking fish migration. In that case heat is not the problem (falling water is spinning the turbines, not steam); the problem is that the fish cannot swim over the barrier posed by the dam. A fish ladder is built to bypass the dam. A fish ladder consists of a number of very small increases in elevation that fish are capable of jumping.

Of course, during migration downstream to the ocean, there are always a few fish that get caught in the turbine blades and are chopped up into little pieces that then float downstream. This is why you are likely to see bald eagles, osprey, and gulls in good numbers feeding in the water just below a dam.

Important Point

Atmospheric CO_2 absorbs Earth's infrared radiation, contributing to global warming. In turn, global warming causes the oceans to become warmer. At higher temperatures, ocean water holds less dissolved CO_2, releasing CO_2 to the atmosphere. As the concentration of atmospheric CO_2 increases, even more global warming occurs.

A third example of the lower solubility of gases at higher temperatures is the effect of temperature on the solubility of carbon dioxide in water. Carbon dioxide is the principal greenhouse gas that is believed to be responsible for global warming. The oceans of the world represent a tremendous sink for CO_2. Carbon dioxide is much more soluble in water than most gases are because CO_2 chemically reacts with water to form carbonic acid (H_2CO_3). The oceans, therefore, remove a large amount of CO_2 from the atmosphere. If the waters of the world's oceans warm, however, then they will absorb less CO_2, and, in fact, will release much of the CO_2 presently dissolved in them back to the atmosphere, further increasing carbon dioxide's greenhouse effect.

The Effect of Pressure on Solubility

The concentration of solids and liquids in solutions tend to be relatively insensitive to changes in pressure. The concentration of gases in liquids, however, varies considerably with changes in pressure. The relationship between gas pressure and solubility is explained by Henry's law.

Henry's Law

Henry's law states that the solubility of a gas in a liquid is directly proportional to the pressure of the gas above the liquid. In other words, if the pressure of the gas is increased, more gas is forced to go into the solution. If the pressure of the gas is decreased, gas tends to escape from the solution. Mathematically, Henry's law is expressed in the following equation:

$$\text{Solubility} = K \cdot \text{Pressure},$$

where K is called the Henry's law constant and has a specific value for each combination of gas and liquid solvent.

Again, a carbonated beverage will be used as an example. CO_2 is forced into the soft drink. During the bottling process, the beverage does not completely fill the bottle. A small space filled with CO_2 is left in the neck of the bottle to maintain a high pressure of CO_2. By Henry's law, that keeps the beverage saturated with CO_2. Remove the bottle cap, and the CO_2 in the neck escapes into the atmosphere. Since the pressure of CO_2 above the liquid has dropped, the solubility of CO_2 decreases and CO_2 begins to escape from the bottle. The soft drink loses its fizz.

The consequences of Henry's law are familiar to scuba divers. At the surface of the water, air pressure is roughly 1 atmosphere. Descend 33 feet and the pressure doubles. Pressure continues to increase with increasing depth. Normally, the nitrogen gas in air is fairly insoluble in water or blood, which is mostly water. If the scuba diver is breathing regular air, however, the increase in pressure increases the solubility of nitrogen in the diver's blood. Should the diver ascend too rapidly, there is insufficient time for the nitrogen to escape into the diver's lungs. Less pressure causes the nitrogen gas bubbles to expand, causing a painful condition called the bends. Not only are the bends extremely painful, they can even be fatal.

One solution is for the diver to ascend gradually, to give the nitrogen time to escape into the lungs again. Applying Henry's law, though, another solution is to breathe a mixture of oxygen and helium instead of the normal mixture of oxygen and nitrogen. Helium is much less soluble in blood than nitrogen is, so very little helium is absorbed into the bloodstream. This should allow the diver to ascend at a faster rate. Of course, breathing helium causes a person to talk like Donald Duck, but scuba divers tend not to do much talking when they are underwater!

The Effect on the Vapor Pressure of a Liquid When a Solute Is Added

All liquids evaporate at least to some extent at any temperature. Evaporation occurs as molecules at the surface of the liquid escape into the gas, or vapor, phase. If the liquid is inside a closed container, molecules in the vapor phase will also be captured at the surface of the liquid and returned to the liquid phase. Eventually, a state of equilibrium will be achieved in which the number of molecules in the vapor phase remains constant. Equilibrium is a dynamic state: molecules are still in motion, and continue to go back and forth between the liquid and vapor states. The number of molecules in each state does not change, although which molecules are in which state is constantly changing.

The pressure exerted by the molecules in the vapor state is called the vapor pressure of the liquid. Only two variables affect the numerical value of a liquid's vapor pressure: the identity of the liquid and its temperature. The identity of the liquid determines the kind of force that exists between molecules of the liquid. Strong intermolecular forces of attraction tend to keep the molecules in the liquid state, so the vapor pressure is low. Weak forces allow more molecules to escape into the vapor state, so the vapor pressure is high. Higher temperatures increase the average kinetic energy of the molecules, which also allow more molecules to escape from the liquid state into the vapor state. Thus, higher temperature means higher vapor pressure; lower temperature means lower vapor pressure.

Dissolving a nonvolatile solute in a liquid also lowers the vapor pressure. We can state with confidence that at any given temperature, the vapor pressure of a solution is less than the vapor pressure of the pure liquid. The reason for this is that evaporation occurs at the surface of the liquid. With solute present, the number of solvent molecules at the surface is less than it would be in the absence of any solute. Fewer molecules means fewer molecules that can escape into the vapor state.

Chem Vocab

A nonvolatile substance is one that does not evaporate, so it contributes negligibly to the solution's vapor pressure.

The decrease in vapor pressure that occurs when a nonvolatile solute is added to a liquid is often called vapor pressure lowering. The extent of vapor pressure lowering depends only on the number of solute particles that are added, and not on the identity of the solute or any other properties like molecular weight or the mass of solute particles. Vapor pressure lowering is one of several properties of solutions that depend only on the number of solute particles that are added. Together, these properties are called colligative properties of solutions. The word *colligative* means that the properties have a common origin—the number of solute particles. (In this discussion we will assume that the solute is not a substance that dissociates into ions when it dissolves. That would increase the number of particles and complicate the calculations.)

The extent of vapor pressure lowering is given by Raoult's law, which states the following:

$$\Delta P = \chi_{solute} \cdot P_{solvent},$$

where ΔP is the decrease in vapor pressure, χ_{solute} is the mole fraction of solute particles, and $P_{solvent}$ is the vapor pressure of pure solvent. For example, the vapor pressure of water is 21 mm Hg at 23°C (73°F). If 0.200 mole of sugar is added to 0.800 mole of water, then the mole fraction of sugar is

$$0.200 \div (0.200 + 0.800) = 0.200.$$

The decrease in vapor pressure is calculated as follows:

$$\Delta P = (0.200) \cdot 21 \text{ mm} = 4.2 \text{ mm}.$$

The vapor pressure of the water is only 17 mm instead of 21 mm.

Calculations of vapor pressure when volatile solutes are added can also be done but are more complicated since a volatile solute contributes to the vapor pressure along with the solvent.

The Effect on the Freezing Point of a Liquid When a Solute Is Added

Again, assume that the solute is nonvolatile and nonionic. The effect of adding a solute to a liquid is to lower the freezing point of the liquid. Since the change in freezing point is independent of the identity of the solute, and depends only on the number of solute particles, it is our second colligative property. The scientific term often used to describe the decrease in freezing point is *freezing point depression*, or just *freezing point lowering*.

The following equation gives the change in freezing point of a solution:

$$\Delta T_f = m \cdot K_f,$$

where m is the molality of the solution and K_f is the freezing point depression constant for the solvent. Table 23.1 shows some values of normal freezing points and freezing point depression constants for a sample of liquids. The values of K_f are all negative to ensure that $\Delta T_f < 0$; in other words, the freezing point of the solution will be lower than the freezing point of the pure solvent.

Table 23.1: Molal Boiling Point and Freezing Point Constants

Substance	T_b(°C)	K_b	T_f(°C)	K_f
water	100	+0.512	0.00	-1.86
acetic acid	118	+3.07	16	-3.9
benzene	80.1	+2.62	5.53	-4.90
carbon tetrachloride	76.8	+5.03	-22.96	-31.8
chloroform	61.7	+3.88	-63.5	-7.3
camphor	208	+5.95	178.4	-37.7
ethanol	78.4	+1.20	-115	-1.99
naphthalene	218	+5.65	80.2	-6.9
phenol	182	+3.56	42	-7.27
diphenyl	254.9	+7.06	+70.6	-8.00
aniline	184.3	+3.69	-5.96	-5.87
stearic acid	383	---	+69.6	-4.50
paradichlorobenzene	174	---	+53.2	-7.10

Notice how small the value of K_f is for water compared to many of the other substances. A small value of K_f means that the freezing point does not change very much when a solute is added. A large value of K_f, as carbon tetrachloride and camphor have, for example, means a very large change in freezing point when a solute is added. The small value of K_f for water is another example of the pronounced effect of hydrogen bonding on water's properties. In Table 23.1, acetic acid and ethanol also exhibit hydrogen bonding, and they also have low values of K_f.

As a sample problem, suppose we dissolve 2.00 grams of phenol (C_6H_6O) in 40.00 grams of paradichlorobenzene. Paradichlorobenzene normally freezes at 53.2°C (127.8°F). Its value of K_f is –7.10°C/m. At what temperature will the solution freeze?

To solve this problem, the first thing we need to do is to calculate the molecular weight of phenol. Given that the formula for phenol is C_6H_6O, we calculate a molecular weight of 94.12 g/mol. Therefore, 2.00 grams of phenol is 0.0212 mole.

Next, we need to change 40.00 grams of paradichlorobenzene to 0.04000 kg. That means that the molality of the solution is

$$0.0212 \text{ mole} \div 0.04000 \text{ kg} = 0.530 \text{ m.}$$

Finally, the following calculation gives us the change in freezing point:

$$\Delta T_f = (0.530 \text{ m}) \, (-7.10°\text{C/m}) = -3.8°\text{C.}$$

Since the normal freezing point is 53.2°C, the solution freezes at 53.2°C – 3.8°C = 49.4°C.

Freezing point depression is the principle behind the use of antifreeze in motor vehicle cooling systems. Water normally freezes at 0°C (32°F). In many places in the United States and Canada, winter temperatures fall well below 0°C. Since water expands as it freezes, a car or truck's engine block and radiator would crack. A 50/50 mixture (by volume) of water and antifreeze, though, provides protection to –34°C (–29°F), which is sufficient for most places that experience below freezing temperatures. If temperatures might get even lower, a mixture of up to two-thirds antifreeze and one-third water provides protection to temperatures of –90°F (–68°C).

The Effect on the Boiling Point of a Liquid When a Solute Is Added

The effect of adding a solute to a liquid is to raise the boiling point of the liquid. Since the change in boiling point is independent of the identity of the solute and depends only on the number of solute particles, it is our third colligative property. The scientific term often used to describe the increase in boiling point is *boiling point elevation*.

The equation for boiling point elevation is analogous to the equation for freezing point depression:

$$\Delta T_b = m \cdot K_b,$$

where K_b is the boiling point elevation constant for the solvent. Table 23.1 shows some values of normal boiling points and boiling point depression constants for a sample of liquids. The values of K_b are all positive to ensure that $\Delta T_b > 0$; in other words, the boiling point of the solution will be higher than the boiling point of the pure liquid. Again, notice how small the value of K_b is for water compared to many of the other substances. This is again due to the effect of hydrogen bonding between water molecules.

Conclusion

The effects of the physical properties of solutions are very much a part of our daily lives, from mixing paint to scuba diving to adding antifreeze to our cars' radiators. From the examples given in this chapter, you should now have a better understanding of how these phenomena work.

CHAPTER 24

The Chemistry of Water

In This Chapter

➤ The polarity of water

➤ Effects of hydrogen bonding

➤ Water's true nature

➤ A look at water as a solvent

Water (H_2O) is our most familiar liquid. It is essential to life as we know it. But how much do most people really know about the physical and chemical properties of water? Water, in fact, is unique in many ways that we usually just take for granted. Water possesses many special properties that result from water's structure on a molecular level.

The Shape of a Water Molecule

We begin with the Lewis electron dot structures for hydrogen and oxygen atoms. These structures are the following: H· and :Ö·. Because hydrogen atoms have only one electron, they can form only one chemical bond. Because oxygen atoms have two unpaired electrons, they tend to form two chemical bonds.

Putting two hydrogen atoms together with one oxygen atom suggests a structure like the following:

$$H - \overset{..}{O} - H,$$

which certainly makes it look like the three atoms lie in a straight line. The whole point of this discussion, however, is that they do not lie in a straight line! If they did, water molecules would be nonpolar and water would be very much like carbon dioxide (CO_2), whereas, in reality, water and carbon dioxide are not very much like each other at all.

Important Point

If water molecules were linear, water would be nonpolar, and life on Earth would not exist.

No, water molecules have a bent shape to them, with the angle between the two OH bonds being about 104.5°. The four pairs of electrons on the oxygen atom are arranged in the shape of a tetrahedron, an arrangement that necessarily pushes the hydrogen atoms out of alignment with the oxygen atom. Since oxygen atoms are more electronegative than hydrogen atoms are, water is a polar molecule. In fact, it is *very* polar—one of the most polar of all common molecules. All of water's other properties derive from its polarity.

The Many Effects of Hydrogen Bonding

Water's polarity means that there is strong hydrogen bonding between water molecules. One result is an unusually high melting point, an unusually high boiling point, and a very wide liquid range for such a small molecule. Most small molecules—CH_4, CO_2, NH_3, H_2S, for example—have such low boiling points that they are gases below 0°C (32°F) at 1 atm pressure. Water, on the other hand, is a solid (ice) under those conditions. Ice does not melt until it reaches 0°C. Once melted, water remains a liquid at temperatures all the way up to 100°C (212°F). (Most substances have much a narrower range of temperature over which they are liquids.)

Important Point

Hydrogen bonding keeps water in the liquid state that is essential to life, and hydrogen bonding is what holds together the strands of atoms in the DNA double helix.

Higher temperatures are necessary to convert solid water into liquid and liquid water into vapor because a lot of energy is required to overcome the forces of attraction caused by hydrogen bonding. So, water's high melting point and boiling point, and water's large range of temperatures for the liquid state are anomalous.

Ice Floats

Another unusual property of water is that liquid water expands as it freezes, whereas most liquids contract as they freeze. Ordinarily, the solid phase of a substance is denser than its liquid phase, so the solid sinks in its own liquid. Ice, on the other hand, floats. It is true that water starting at room temperature does contract as it cools down. However, the contraction stops at 4°C (39°F), at which point a sample of water reaches its minimum volume (and maximum density).

Below 4°C, the water molecules begin to rotate so that their relative orientations maximize the amount of hydrogen bonding that can occur. As they do so, the molecules begin moving apart again just slightly, so that by the time they reach 0°C and actually freeze, the ice that forms is less dense than the liquid water around it and the ice floats. So, the fact that ice floats is anomalous. (In fact, for the solid state of water to float is so extremely unusual that the only other reasonably common example is the element bismuth.)

Important Point

People just take it for granted that ice floats. That is a very unusual phenomenon, however, and rare among elements and compounds.

Capillary Action

Another phenomenon associated with water that is due to hydrogen bonding is called capillary action. An example of capillary action is what happens when a thin glass tube is inserted into water; water molecules hydrogen bond to the oxygen atoms in the glass (glass is SiO_2), causing water to slightly "creep up" the walls of the glass. The result is a slight curvature to the upper surface of the water that is called a meniscus. Nonpolar molecules do not do this: gasoline, liquid mercury, and oils do not form a meniscus. So, capillary action is anomalous.

Rate of Evaporation

Pour any liquid into a shallow dish and the liquid begins to evaporate. Picture, for example, a shallow dish of gasoline and a shallow dish of water that are both placed in the sunshine. Which liquid evaporates more quickly? The gasoline, of course. Gasoline molecules are nonpolar, so the forces of attraction between molecules are very weak (and are weaker the smaller the molecules) and gasoline easily goes from the liquid phase to the vapor phase. On the other hand, water evaporates much more slowly because the strong force of hydrogen bonding tends to keep the molecules in the liquid phase. So, a slow rate of evaporation is anomalous.

Vapor Pressure of Water

Any nonpolar substance with molecules as small as water molecules would be a gas under normal conditions of 1 atm pressure and 20°C. Larger nonpolar molecules might be a liquid under normal conditions, but they would be a volatile liquid, meaning they would easily evaporate and have a high vapor pressure. Gasoline is an example of a mixture of medium size nonpolar hydrocarbons. Gasoline vaporizes easily (which, of course, it needs to do in a vehicle's spark plug chamber), leading to high vapor pressures. On the other hand, water is relatively nonvolatile and has a relatively low vapor pressure at ordinary temperatures. Water's low vapor pressure is again a result of hydrogen bonding. Because the molecules are so strongly attracted to each other in the liquid state, water does not evaporate easily.

Important Point

Other substances that exhibit hydrogen bonding may also exhibit low vapor pressures. Antifreeze contains ethylene glycol because its molecules hydrogen bond to each other making ethylene glycol nonvolatile. In addition, ethylene glycol and water molecules hydrogen bond to each other, making the two liquids miscible.

Water's Specific Heat

When heat is added to most liquids, the temperature of the liquid readily increases. Not so with water, however. We express the amount of heat required to increase the temperature of 1 gram of a substance 1°C as the specific heat of the substance. The units are J/g°C. Most common metals have a specific heat in the range of 0.1–1.0 J/g°C (see Table 24.1). Common liquids have a specific heat in the range of 0.8–2.3 J/g°C. But water has an exceptionally high specific heat—4.18 J/g°C. Once again, the difference is due to hydrogen bonding between water molecules. Whereas most other substances have to absorb only a little bit of heat to warm up, the hydrogen bonds are capable of absorbing a relatively large amount of heat with only a small rise in temperature.

Table 24.1: Specific Heats of Common Substances

Substance	Specific Heat (Joules/g˚C)
Water (liquid)	4.18
Water (ice)	2.10
Ethyl alcohol	2.46
Aluminum	0.900
Iron	0.444
Copper	0.385
Silver	0.240
Gold	0.129
Lead	0.160

You can see the difference when walking in bare feet on a sandy beach on a hot day. The specific heat of sand is about 20 percent that of water, so the sand warms up considerably (sometimes to the point of discomfort!), while the temperature of the water is cooler and very pleasant.

The effect of water's high specific heat is also known to owners of greenhouses. On cold nights, placing a tub of water in a greenhouse keeps the interior of the greenhouse warmer even if some of the water freezes. Heat that was stored in the water is released to the air in the greenhouse during the night, keeping the air warmer.

A car's cooling system also operates off the principle that the water (and coolant fluid) that is circulating through the engine draws heat away from the engine. As the heated water percolates down the core of the radiator, air drawn through the radiator by the fan absorbs the heat and releases it to the atmosphere. The relatively cooler water then circulates through the engine again.

This property of water has a huge effect on climate. Draw a line at latitude 45°N across the continental United States. This line will roughly pass through Seattle, Washington; Pocatello, Idaho; Missoula, Montana; Fargo, North Dakota; and Minneapolis, Minnesota. Of all these cities, Seattle has the mildest climate—relatively cool summers and warm winters. The other cities are known for their extremely cold winter climates. The difference results from Seattle's location on Puget Sound. The waters of Puget Sound absorb heat from the Sun during the summer, keeping the air temperature relatively cool, and release heat during the winter, keeping the temperature relatively warm. Pocatello, Missoula, Fargo, and Minneapolis have no such large bodies of water to regulate temperature, so their climate fluctuations between summer and winter are much more extreme. Again, water's relatively high specific heat is anomalous.

Important Point

Water's high specific heat allows water to absorb heat without much of a temperature change. The presence of large bodies of water—lakes, bays, estuaries—helps regulate climate.

Chem Vocab

Latent means "hidden," or "not evident." Since temperature does not change during phase changes, it is not evident that heat is being absorbed when a solid melts or a liquid vaporizes.

Latent Heats of Fusion and Vaporization

The amount of heat required to melt a substance in the solid state is called its latent heat of fusion, and the amount of heat required to vaporize a liquid is called the substance's latent heat of vaporization. Once again, water has relatively high latent heats of fusion (333 J/g at 0ºC) and vaporization (2257 J/g at 100ºC), which are especially high for such a small molecule. This is why it takes a lot of heat to boil water, and why water boils more easily in a pot with the lid on than in an open pot.

Important Point

Remember the saying "a watched pot never boils"? Well, this is why: without the lid on, the heat just escapes rather than staying to raise the water's temperature. This is also why steam can scald a person. All that heat that had to be absorbed to produce the steam is still present in the steam in the form of kinetic energy. If that steam touches your skin, all that extra energy is transferred back to your skin as heat.

Water's high heat of vaporization is the principle behind the cooling effects of perspiring. For water on our skin to evaporate, the water has to absorb heat from the skin, which in doing so, cools down the skin. However, once again, water's high heats of fusion and vaporization are anomalous.

Surface Tension

Surface tension is the force between molecules at the surface of a liquid that tends to hold the molecules on the surface. (In physics, the word *tension* refers to a kind of force.) The relative amount of surface tension in a liquid is determined directly by the strengths of the forces of attractions between molecules of that liquid. Nonpolar liquids have almost no surface tension; water has a very high surface tension because of hydrogen bonding.

Chem Vocab

Surface tension is the uneven force on molecules at the surface of a liquid that tends to pull molecules back towards the liquid's interior.

The surface tension of water is responsible for the spherical shape of rain drops and the ability of water to form beads on a freshly waxed surface. Wax is a nonpolar substance, so there is almost no force of attraction between water molecules and wax molecules. The water molecules are much more strongly attracted to other water molecules, hence the beads. Again, water's high surface tension is anomalous.

The Shape of Snowflakes

Hydrogen bonding is responsible for the geometrical structures of snowflakes. Like ice, a snowflake is another form of solid water in which hydrogen bonding between water molecules is optimized. Under the conditions of temperature and humidity at which snowflakes form, hydrogen bonding leads to snowflakes' dazzling structures. The six-sided figures that snowflakes form are just six tetrahedrons forming a hexagon.

The Discovery of Water's True Nature

Ancient Greek philosophers believed that matter is composed of four fundamental elements: earth, air, fire, and water. It was two thousand years, however, before the elements hydrogen and oxygen were discovered and people realized that water is not a single element, but a combination of hydrogen and oxygen. The name *hydrogen*, in fact, means "water forming," reflecting the fact that water is made in part from hydrogen.

Early Greek philosophers also believed that matter was composed of tiny, indivisible particles that they called atoms, from a Greek word that means "indivisible." Aristotle, one of antiquity's most influential philosophers in areas of natural science, discounted the existence of atoms and asserted that matter, in principle at least, can be divided over and over again indefinitely.

With the discoveries of hydrogen, oxygen, nitrogen, and a quantitative understanding of the conservation of matter in ordinary chemical reactions, by the year 1810 chemists were beginning to believe again in the existence of atoms. Since hydrogen atoms, oxygen atoms, and water molecules are involved in a great number of chemical reactions, understanding the atomic nature of matter lead to a firmer understanding of the difference between elements and compounds. You need to realize that all study of elements and compounds was very difficult for chemists in the early 1800s. For many years, chemists even debated whether the formula for water should be HO or H_2O.

This understanding placed chemistry on a much more solid footing as a true science. Much of this advancement can be attributed to chemists' understanding of water and the recognition of the many properties of water we have just discussed (although the discovery of hydrogen bonding did not occur until the first half of the twentieth century).

Chem Vocab

Heavy water means that it is heavier than ordinary water. A molecule of ordinary water that contains $_1^1H$ atoms has a mass of 18 amu. A molecule of heavy water that contains deuterium atoms, $_1^2H$, has a mass of 20 amu.

Heavy Water

Another discovery of water's true nature also occurred in the twentieth century—the discovery of heavy water. Ordinarily, a hydrogen atom consists of one proton (in the nucleus of the atom) and an electron (outside the nucleus), but no neutrons. In 1934, the American physical chemist Harold Urey (1893–1981) discovered a heavier isotope of hydrogen that contains one proton, one electron, and one neutron. Since protons and neutrons have about the same mass, and electrons are very, very light in comparison, atoms of heavy hydrogen (deuterium) are about twice as heavy as ordinary hydrogen. The atomic weight of ordinary hydrogen is 1 g/mol and the atomic weight of oxygen is 16 g/mol.) Therefore, water that contains ordinary hydrogen atoms has a molecular weight of 18 g/mol. Water that contains deuterium atoms is called heavy water, has a higher molecular weight (about 20 g/mol), and is sometimes written as D_2O.

Heavy water has a few properties that are different from the properties of ordinary water. One difference is that, at 1 atm pressure, heavy water freezes at 3.8°C (38.8°F) and boils at

101.4°C (214.5°F). Another difference is that living organisms cannot use D_2O. While H_2O is absolutely essential to all forms of life as we know them, all living organisms would die if only given D_2O. Heavy water does exist naturally on Earth, but at a relatively tiny proportion compared to ordinary water.

Nevertheless, the oceans of the world are such huge reservoirs of water that the amount of D_2O is significant. Nuclear fusion reactors would use deuterium as a fuel (along with tritium, the still heavier isotope of hydrogen), but the technological obstacles to the successful development of commercial fusion reactors have never been overcome. If these obstacles were overcome in the future, there is enough deuterium in the waters of the world's oceans to supply the needs of fusion reactors for thousands of years.

Water as a Solvent

Salts, which are ionic compounds, are insoluble in most liquids, especially nonpolar liquids like gasoline. Many salts, however, are very soluble in water. Because of water's polar nature, the oxygen end of a water molecule carries a slight negative charge and the hydrogen end of the molecule carries a slight positive charge. It is true that the positive and negative ions in salt crystals have a strong attraction for each other. Very often, however, the attraction of the positive ions to the oxygen end of water molecules and the attraction of negative ions to the hydrogen end of water molecules is stronger than the forces of attraction between the ions themselves. When that happens, the salt dissolves and the ions are hydrated. This is especially true if the salts contain one of the following ions: Na^+, K^+, NO_3^-, or $C_2H_3O_2^-$.

An important point to make is that water is a relatively inert solvent; in other words, the properties of water do not really change when another substance—solid, liquid, or gas—dissolves in it. This is especially vital to living organisms. Aqueous solutions (blood, for example), are the transport media for an organism's essential inorganic and organic substances. These substances can flow to different parts of the body unchanged. Water itself also tends to remain unchanged in the process. Lakes, rivers, and streams store and transport a large quantity of inorganic and organic substances; again, the water in those bodies does not change in the process.

Aquatic Ecosystems

Substances that are soluble in water often are in the form of ionic compounds, consisting of positive and negative ions. Common salts, such as ordinary table salt (sodium

Chem Vocab

Salinity is the degree of saltiness of seawater. In the open ocean the concentration of dissolved salts is about 30 g/L of seawater.

chloride, or NaCl), dissolve in lakes and rivers and are washed to the oceans, giving ocean water its salinity, or saltiness. (Oceans contain other salts, too, but NaCl is present in greatest concentration.) At the concentrations of Na^+ and Cl^- ions normally found in aquatic systems, NaCl is soluble. As the water evaporates, however, salt crystals begin to form. Underground salt deposits, such as those under the city of Detroit, Michigan, were left behind when the land once submerged under the ocean rose and the water evaporated.

Salt crystals are a familiar sight along the shores of salt lakes such as the Great Salt Lake in Utah, the Salton Sea in southern California, Mono Lake in eastern California, and the Dead Sea in Israel. After swimming in the ocean, you may have licked your lips and tasted the salt that had precipitated on them. Commercial table salt companies flood shallow basins with salt water, allow the water to evaporate, and dredge the salt that is left behind.

Carbon dioxide (CO_2) is slightly soluble in water (as shown below), giving the familiar carbonic acid (H_2CO_3) found in soft drinks. In aquatic systems, H_2CO_3 dissociates into bicarbonate ions (HCO_3^-) and carbonate ions (CO_3^{2-}), as shown by the following equations:

➤ $CO_2\ (aq) + H_2O\ (l) \leftrightarrow H_2CO_3\ (aq)$

➤ $H_2CO_3\ (aq) \leftrightarrow HCO_3^-\ (aq) + H^+\ (aq)$

➤ $HCO_3^-\ (aq) \leftrightarrow CO_3^{2-}\ (aq) + H^+\ (aq)$.

These reactions can take place in both directions depending upon temperature, acidity, and the presence of other ions. Notice that when hydrogen ions (H^+) are formed, they tend to make natural aquatic ecosystems slightly acidic.

Limestone (calcium carbonate, $CaCO_3$) is a common sedimentary rock that was laid down eons ago on ocean floors, mostly from the remains of decomposed marine organisms. Subsequent uplift of the land brought limestone above sea level. Carbon dioxide dissolved in rainwater undergoes the reactions shown above. The natural acidity of rainwater lends itself to slightly dissolving limestone, releasing calcium ions (Ca^{2+}) into the environment, as shown:

$$CaCO_3\ (s) + 2\ H^+\ (aq) \rightarrow Ca^{2+}\ (aq) + CO_2\ (g) + H_2O\ (l).$$

Those calcium ions may flow in surface water and in ground water. Once again, given the right conditions of temperature, acidity, and the presence of other ions, $CaCO_3$ may precipitate again, sometimes forming striking formations. Examples include the stalactites growing on the ceilings and stalagmites growing on the floors of caverns such as Carlsbad Caverns in southern New Mexico, Mammoth Caves in Kentucky, Grand Caverns in Virginia, Wind Cave in western South Dakota, and Karchner Caverns in southern Arizona. Another example is the travertine cliffs of Havasu and Mooney Falls on the Havasupai Indian Reservation at the western end of Grand Canyon, Arizona.

Conclusion

Water is a substance that is essential to life and that has numerous important properties. Hopefully, this chapter has given you an appreciation for what those properties are.

PART FIVE

Chemistry in the Environment

CHAPTER 25

 **Chemical Kinetics**

In This Chapter

- ➤ Collision theory
- ➤ Transition state theory
- ➤ Factors that affect the rate of a chemical reaction

Chemists seek to understand the details of chemical reactions:

> ➤ How atoms and molecules behave as they come together to form products

> ➤ What rates and mechanisms are characteristic of reactions

> ➤ How to control the rates at which products are formed

These topics are the subject of that branch of chemistry called chemical kinetics.

Collision Theory and the Three Conditions for Products to Form

There are three conditions that are necessary for reactants to be converted into products. Collision theory explains what these conditions are.

The first condition is that molecules have to collide with each other. If they do not collide, no reaction will take place.

Important Point

Collision theory tells us that three conditions must be satisfied for products to form:

➤ Reactant species must collide with each other.

➤ Molecules have to hit each other hard enough to break the bonds in the reacting species.

➤ Molecules must come together with the correct relative orientation to each other.

The second condition is that the force of the impact between two colliding molecules must be hard enough that the atoms can rearrange and form the different combinations of bonds found in the product molecules. If impact is not hard enough, the collision is a "dud" and no products are formed.

The third condition is that the molecules must also collide at the correct relative orientation to each other. If molecules do not hit each other at the correct places, nothing will happen. For example, consider the reaction between carbon monoxide and oxygen gas to form carbon dioxide:

$$2\ CO\ (g) + O_2\ (g) \rightarrow 2\ CO_2\ (g).$$

The arrangement of atoms in a CO_2 molecules is $O = C = O$. In order for a CO molecule to become a CO_2 molecule, it should be clear that an O_2 molecule has to hit the carbon atom on the CO molecule. Hitting the oxygen atom on CO has no effect because we are not trying to make $C - O - O$. Therefore, the relative orientation of the reacting species is crucial.

Chemical Reactions and Transition State Theory

Another theory that also helps us understand how chemical reactions take place is called transition state theory (TST), the development of which is attributed primarily to the American chemist Henry Eyring (1901–81), who spent most of his career at the University of Utah in Salt Lake City. In TST, we picture an energy barrier that the reacting molecules must pass over before they can become products. Consider an analogy.

Suppose you are standing in front of the goal posts on a football field and trying to kick the ball over the horizontal bar between the posts (you are trying to kick a field goal). The kinetic energy your kick imparts to the ball has to be strong enough to get the ball over the bar. If you kick the ball with insufficient energy, you will not score any points. In a chemical reaction, we call the minimum energy necessary to reach the transition state the activation energy for the reaction.

Replace the football with molecules. The molecules are all at the same temperature, but they are moving at different speeds and, therefore, with different kinetic energies. The activation energy for the reaction is the minimum kinetic energy that allows the impact of collision to be hard enough for the molecules to reach the reaction's transition state, the highest point on the energy curve. Relatively few molecules possess energy greater than the activation energy. Only those molecules with energy greater than or equal to the activation energy pass through the transition state to become products.

Two outcomes can result from the transition state. The transition state species can just break apart into the original molecules with no net change, or it can continue on and form the products of the reaction.

A reasonable question to ask is if it's possible for the products of a reaction, once they have formed, to collide with each other, go back through the transition state, and reform the reactants with which we started originally? The answer is yes.

Chem Vocab

The transition state is the molecular intermediate occurring at the highest point on the energy level diagram for a chemical reaction.

Important Point

Most chemical reactions are reversible. Whether a reaction is proceeding from reactants to products or from products back to reactants, the transition state is the same in both directions.

Important Point

In an endothermic process, the activation energy of the reaction is higher in the forward direction than it is in the reverse direction.

In an exothermic process, the activation energy is higher in the reverse direction than in the forward direction.

First of all, in theory at least, all chemical reactions are reversible. The transition state has the same structure whether it is formed from collisions of reactant molecules or from collisions of product molecules. The requirements for products going back to reactants (the reverse direction) are the same as they are for the forward direction. What most often is the case, however, is that the activation energy is different in the two directions.

The Five Factors that Affect the Rate of a Chemical Reaction

There are five factors that affect how fast a chemical reaction will take place:

➤ The nature of the reactants: It makes sense that various substances react with other substances differently. For example, metals dissolve in acids at different speeds. The exhaust gases from an automobile enter the atmosphere where they react with other gases in the atmosphere at different rates.

➤ Surface area of solids: Some chemical reactions take place on solid surfaces. The more surface area that is available, the more rapidly the reaction can occur. Finely divided solids have more surface area than bulk solids, so reactions on finely divided solids occur more rapidly.

➤ Concentration or pressure of reacting species: A higher concentration of ions or molecules dissolved in solution means that they will collide with each other more frequently, which results in products being formed at a faster rate. Gases at a higher pressure also means that molecules can collide with each other more frequently.

➤ Temperature: The vast majority of chemical reactions proceed at a faster rate at higher temperatures.

➤ The presence of catalysts: A catalyst is defined as a substance that speeds up a chemical reaction without itself being consumed by the reaction. When the reaction is finished—when all of the reactant molecules that are going to react have, in fact, reacted—the catalyst will again be in the same state and quantity as it was at the beginning of the reaction.

Let us look at each of these factors in turn.

Important Point

There are five factors that affect the rate, or speed, with which a chemical reaction takes place: the nature of the reactants, the surface area of solids, the concentration of reactants, temperature, and the presence of catalysts.

The Rates of Reactions Depend on the Nature of the Reactants

There are a lot of chemicals that we can mix together and nothing at all will happen. For example, when we mix salt and sugar, no chemical reaction takes place between them. Sprinkle salt in a flame and it does not burn. Sprinkle sugar in a flame, however, and the sugar will catch on fire, that is, it will undergo combustion. Now ignite a lump of charcoal, a similar-sized block of wood, a piece of cloth saturated with rubbing alcohol, and a piece of hamburger. If we supply enough heat they will all burn, but we will see them being consumed at different speeds.

If we collected the carbon dioxide and water vapor produced by each reaction in separate bottles, the rate at which the pressure of carbon dioxide and water vapor increases would be different for each bottle. Now, you may be curious as to which of these examples would burn faster. There really is not a good way to predict what the results would be. That is why scientists do experiments!

The Rates of Reaction Depends on the Surface Area of Solids

You probably already know from your own experiences that the surface area of a solid determines the rate of reaction. At some time in your life you may have built a campfire or started a fire in a fireplace or lit charcoal in a grill on fire. If you built a campfire, for example, you probably discovered that small pieces of wood (tinder) catch on fire more readily than a log. So it is better to start fires with small twigs than with logs. As the twigs catch on fire, you can gradually add larger and larger pieces of wood until finally a decent-sized log can be placed on the fire. You let that burn down until there is a good set of red-hot coals, and finally get out the marshmallows, graham crackers, and chocolate bars and make s'mores!

If a chemical reaction takes place on the surface of a solid, the more surface area there is, the more molecules that can react at one time. If more molecules are reacting per unit time, then more products are also being formed per unit time, which is what we mean by saying the reaction is going faster. Twigs have a larger surface area than logs do, so twigs catch on fire faster than logs do. (Dryer lint has even more surface area, so it catches on fire even faster!)

Important Point

When a chemical reaction takes place at the surface of a solid, the more surface area the solid has, the faster the reaction will take place. Many tiny pieces of a substance will have a much larger surface area than one big lump of the substance.

The Effect of Concentration

The more people there are in a crowded room, the greater is the likelihood that people will bump into each other. In the same way, if we increase the number, or concentration, of the reacting species, it should make sense that molecules will collide with each other more often. In the case of gases, since pressure increases with concentration, we can also say that gases react faster at higher pressures. In either case, collision theory tells us that an increase in the frequency of collisions results in products being formed at a faster rate.

An example would be taking Tums for an upset stomach. Tums are made of $CaCO_3$ and react with acid to generate CO_2 gas (see Chapter 9). If the concentration of stomach acid (HCl) is low, CO_2 will be generated very slowly and you may not even notice it. If the concentration of stomach acid is high, however, CO_2 will be generated rapidly, the result probably being a loud belch as excess CO_2 escapes from your digestive system.

The Effect of Temperature

Molecules travel faster at higher temperatures, so at higher temperatures we expect two results:

➤ Molecules collide more frequently

➤ More molecules have energy exceeding the activation energy for the reaction. The force of collisions is greater than it would be at a lower temperature, and a greater number of collisions will result in the formation of products. Consequently, at higher temperatures reactions proceed more rapidly.

Important Point

A good rule of thumb is that raising the temperature of a system by 10°C roughly doubles the speed at which the reaction takes place.

Any molecule that has kinetic energy equal to or in excess of the activation energy for the reaction, can potentially become products. But even then remember that molecules must come together with the proper orientation.

It should be clear that forming products isn't easy! Recall from collision theory that the vast majority of molecules never collide, produce collisions that are duds, or do not come together correctly in the first place. At any one time only a very tiny fraction of the reactants are actually forming products.

But this is good! Many natural processes need to proceed at a relatively slow rate—examples include the digestion of food, the absorption of oxygen in the lungs, the metabolic processes inside cells, and combustion reactions. If reactions did not take place relatively slowly, chemical processes would take place at such runaway rates that living organisms would overheat and die, fires could not be extinguished, food would burn to a crisp without cooking first, and the atmosphere would be rapidly depleted of all of its oxygen and nitrogen.

Chemical Reactions Take Place More Quickly in the Presence of a Catalyst

Chem Vocab

Catalysts are chemical elements or compounds that speed up a chemical reaction but are not themselves changed by the reaction, either in quantity or identity. Since the same amount of catalyst is present when the reaction reaches equilibrium as was present initially, the chemical formula for the catalyst is not included when writing the balanced chemical equation for the reaction. Nor do catalysts have any effect on the state of equilibrium itself since the catalyst speeds up both the forward and the reverse reactions.

To understand how a catalyst works, consider again the energy barrier faced by the reactant molecules. A catalyst provides the reactants with an alternative path to get to the products, a path that has a lower activation energy. By lowering the energy barrier, more molecules have kinetic energy in excess of the minimum energy necessary to go over the barrier, so the reaction takes place more quickly. But remember, that also lowers the energy barrier in the reverse direction, so products also can form reactants more easily. Because the speed of the reaction is increased in both directions, the final state of equilibrium that is reached is unchanged.

Important Point

Catalysts do not change the relative distribution of reactants and products at equilibrium; they only decrease the time it takes for the system to reach a state of equilibrium.

Important Point

Catalysts have several important properties:

> ➤ A catalyst is not permanently changed during the course of the reaction it is catalyzing.

> ➤ A catalyst can be used again and again to catalyze the reaction between more reactants.

> ➤ A catalyst cannot make a reaction take place that would not occur in the absence of the catalyst.

Let us see how catalysts work. There are two kinds of catalysts: heterogeneous and homogeneous. A heterogeneous catalyst is in a different phase than the reactants are in. The most common example is a solid catalyst such as the catalytic converter in a motor vehicle, which converts harmful gases in the vehicle's exhaust to carbon dioxide, water, and nitrogen (see Chapter 27).

In this case the catalyst is a solid (usually a platinum-group metal) and the reacting species are gases. The function of the catalytic surface is to momentarily hold the exhaust gases in place until they have time to react. Otherwise, the exhaust gases would just pass out through the tailpipe without undergoing any change. What's important, though, is that metal used in the catalytic converter is not changed. It can be used for years to clean a vehicle's exhaust of gases that would be pollutants in the atmosphere.

Solid catalysts find many applications in industry. An important example is found in the Haber process (named for its inventor, Fritz Haber [1868–1934]), in which nitrogen gas (N_2) and hydrogen gas (H_2) are combined to form ammonia (NH_3), an important agricultural fertilizer. The catalyst used in the Haber process is solid iron oxide.

In the case of homogeneous catalysts, the catalyst is in the same phase as the reactants. Sometimes the catalysts may be dissolved in solution along with the chemicals with which they are interacting. Other times, catalysts may be gaseous substances that are part of a mixture of gases. This is often the case in reactions that take place in the atmosphere, although the solid surfaces of dust particles and ice crystals can also act as heterogeneous catalysts. We will see examples of homogeneous catalysis in our discussion of chemicals that are harmful to stratospheric ozone.

In the case of homogeneous catalysis, the catalytic molecules usually are converted into something else during one step of the reaction but are regenerated at the end of the reaction. Therefore, homogeneous catalysts can also be used over and over again to catalyze the reactions between more molecules.

Conclusion

Chemical changes are taking place all the time all around us and inside of us. Understanding how to control the rate at which chemical changes occur has been an important achievement of modern chemistry.

CHAPTER 26

The Composition of the Atmosphere

In This Chapter

➤ Gases in the atmosphere

➤ The regions of the atmosphere

➤ The relationship between altitude and atmospheric pressure

➤ Chemistry of the atmosphere

The atmosphere is a homogeneous mixture of numerous gases. Except for its water content—which varies from 1 to 4 percent of the gases present—the relative concentration of the atmosphere's gases is remarkably constant both with location on Earth's surface and with increases in elevation. An air pollution episode or a volcanic eruption may result in a temporary increase in one or two gases, but the concentration of those gases normally revert to their usual equilibrium concentration. A particularly violent volcanic eruption is an exception because it may spew gases to such extreme elevations that it can take years for them to settle back down again. The eruption of Krakatoa in the South Pacific in 1883 is a notable example. Its gaseous emissions—especially oxides of sulfur—altered global climate for several years after it erupted.

Historical Note

Krakatoa is still an active volcano. In recent years there has been enough seismic activity to make volcanologists (scientists who study volcanoes) believe that Krakatoa is due any time for another eruption, one that could rival the 1883 explosion in severity.

The Relative Concentration of Gases Found in the Atmosphere

Disregarding exceptional circumstances, we can list the relative concentration of gases found in the atmosphere (see table 26.1). Excluding water vapor, we see that only three gases—N_2, O_2, and Ar—make up almost 100 percent of the gases in the atmosphere. Carbon dioxide is the next most abundant gas, but even its concentration is relatively tiny compared to the first three. The concentrations of any other gases are so small that they are usually expressed in units of parts per million (ppm) or parts per billion (ppb), where a part per million means only one molecule of that gas per million air molecules, and a part per billion means only one molecule of a gas per billion air molecules.

Table 26.1: The Composition of the Dry Atmosphere

Major Constituents	Percent by Volume (rounded)
Nitrogen (N_2)	78.08%
Oxygen (O_2)	20.95%
Argon (Ar)	0.934%
Carbon Dioxide (CO_2)	0.036%
Total	**≈ 100%**
Minor Constituents	**Parts per Million (ppm)**
Neon (Ne)	18
Helium (He)	5
Methane (CH_4)	1.7
Krypton (Kr)	1.1
Nitrous Oxide (N_2O)	0.3
Hydrogen (H_2)	0.5
Xenon (Xe)	0.09
Trace Gases	**Concentrations are variable**
Ozone (O_3)	< 0.1 ppm
Sulfur Dioxide (SO_2)	< 0.1 ppm
Sulfur Trioxide (SO_3)	< 0.1 ppm
Nitric Oxide (NO)	< 0.1 ppm
Nitrogen Dioxide (NO_2)	< 0.1 ppm
Ammonia (NH_3)	< 0.1 ppm
Water vapor	≈ 1–4%

Regions of the Atmosphere

As anyone who has traveled from sea level to the top of a mountain knows, the atmosphere becomes thinner with increasing elevation, which means that atmospheric pressure and density both decrease with increasing elevation. (The temperature also decreases with elevation.)

For our purposes we can divide the atmosphere into three regions based on elevation. In order from Earth's surface to the highest elevation, these three regions are the troposphere, the stratosphere, and the ionosphere.

Chem Vocab

Air pressure at sea level is about 1 atm. 1 atm = 760 mm Hg or 29.92 inches Hg.

The units used to measure atmospheric pressure are as follows:

➤ Atmospheres (atm)

➤ Millimeters of mercury (mm Hg)

➤ Inches of mercury (in Hg)

➤ Bars or millibars (mbar)

The conversions are 1 atm = 760 mm Hg = 29.92 in Hg = 1.013 mbar.

On average, atmospheric pressure is 1 atm at sea level (hence the unit atmospheres), or 760 mm Hg. At 5,000 feet (1,500 meters), atmospheric pressure drops to 630 mm Hg, and at 10,000 feet (3,000 meters), atmospheric pressure drops to 520 mm Hg. At the top of Mt. Everest in Nepal, which is at an elevation of 29,035 feet (8,850 meters), the pressure is only 250 mm Hg. The summit of Mt. Everest is poking into the stratosphere.

The air is much colder on top of a 14,000-foot (4,270-meter) peak in Colorado than it is at mile-high Denver, Colorado (1 mile is 5,280 feet, or 1,609 meters), which in turn is usually cooler than the Great Plains, which is much closer to sea level.

The troposphere is where most of the water vapor is located, where the winds are, and where weather takes place. Consequently, the air in the troposphere tends to be well mixed. Even episodes of high levels of air pollutants in cities like Los Angeles, California, last only a few days before winds disperse the pollutants and restore the level of pollutants to their background levels.

Important Point

The regions of the atmosphere from ground level upward are the troposphere, the stratosphere, and the ionosphere.

Atmospheric pressure is so low in the stratosphere (only 1–10 mm Hg) that for all practical purposes the stratosphere is almost a vacuum. The stratosphere is significantly colder than the troposphere. It is so cold that liquid water does not exist there; water exists only as ice crystals. Everyone has seen the result of this by having observed jet trails from aircraft flying through the stratosphere, typically at elevations of 30,000 to 35,000 feet (9,140 to 10,670 meters).

In the troposphere, the combustion of an aircraft's fuel results in carbon dioxide and water vapor in its exhaust. In the stratosphere, however, the water vapor instantly freezes, the result being a jet trail of ice crystals following the aircraft. The stratosphere is where ozone gas (O_3) accumulates. Ozone's concentration is exceptionally low, but its presence is still crucial in that it absorbs much of the harmful ultraviolet (UV) radiation from the Sun and prevents that radiation from reaching Earth's surface. By absorbing UV radiation, ozone causes the stratosphere to be warmer than it would be otherwise. It is still colder than the troposphere, but warmer than would be predicted by extrapolating the rate of decrease in temperature with elevation that occurs in the troposphere.

The ionosphere is yet higher than the stratosphere. As the name suggests, gases in the ionosphere tend to be in the form of ions. For our purposes, we will pass over the chemistry of the ionosphere. We do note, however, that the ionosphere plays a role in climate control.

Chemistry of Earth's Unique Atmosphere

Consider first the natural atmosphere. In this context the term *natural* refers to an atmosphere untouched by human activity. That does not mean, though, that before there were humans there was no air pollution. Pollution can also have natural causes; we just were not around to experience it.

Important Point

Not all air pollution is anthropogenic. Volcanoes spew noxious gases (especially SO_2) into the atmosphere. Decaying vegetation in marshes and swamps emit methane (CH_4) and hydrogen sulfide (H_2S) into the atmosphere. Violent winds can kick up dust and sand storms.

Before humans came on the scene, there were many gases in Earth's atmosphere: N_2, O_2, Ar, water vapor, CO_2, CO, NO, NO_2, SO_2, SO_3, O_3, HNO_3, H_2SO_4, He, Ne, Kr, Xe, Rn, and many organic compounds that resulted from the presence of living organisms on Earth.

Some of these gases—CO, NO_2, SO_2, HNO_3, and H_2SO_4 especially—are classified from a human point of view as air pollutants. In addition, even a natural atmosphere contains soot, particulate matter, and aerosols.

Earth's atmosphere is unique in our solar system. Other planets and some moons have an atmosphere, but Earth's atmosphere is the only one of which we are aware that can support life as we know it. (Life may exist elsewhere in the solar system—and the search for extraterrestrial life is part of many space missions—but the fact remains that the presence of life outside of Earth's biosphere has not been discovered.)

The principal gases in Earth's atmosphere are as follows:

➤ Nitrogen

➤ Oxygen

➤ Water vapor

➤ Argon

➤ Carbon dioxide

Because the amount of water vapor in the atmosphere is so variable, most discussions of the atmosphere's composition exclude water vapor. This chapter, therefore, will be about gases other than water vapor that are found in Earth's atmosphere.

Nitrogen

The gas present in the greatest quantity in the atmosphere is nitrogen, as N_2 molecules. In fact, nitrogen comprises 78 percent (almost four-fifths) of the atmosphere by volume, or number of molecules. Therefore, N_2 is the solvent, and all other gases are considered to be solutes. The reason that N_2 is present in such high concentration is due to the relative nonreactivity of the N_2 molecule. The two nitrogen atoms are held together by a triple bond that is one of the strongest chemical bonds found in nature.

Important Point

The triple bond in a nitrogen molecule, depicted :N≡N:, is one of the strongest chemical bonds found in nature. It is responsible for the relative chemical inertness of nitrogen gas.

As a general rule a lot of energy is required to split a nitrogen molecule into its two atoms. One source of such energy is a lightning bolt. Another source of that much energy is the ignition of the air-fuel mixture that occurs inside an internal combustion engine. When a spark plug ignites, the intention is to combine gasoline vapors with the oxygen present in air. The spark causes such a high temperature, however, that O_2 molecules also react with some of the N_2 molecules producing NO and NO_2 (NO_x) that is tested for in smog checks of motor vehicles.

Neither of these processes, however, is sufficient to make nitrogen available to plants. Nitrogen is the number one nutrient required for plant growth. For nitrogen to become available to plants requires the presence of nitrogen-fixing bacteria attached to the roots of plants, living in a symbiotic relationship with the plant. In this case, high temperatures are not required to split the nitrogen-nitrogen triple bond. Instead, this is a catalytic process. Catalysts allow chemical reactions to occur at lower temperatures than would otherwise be required. Not all plants, however, are capable of fixing nitrogen.

Oxygen

The gas present in second greatest abundance in the atmosphere is oxygen, as O_2 molecules. Oxygen gas comprises 21 percent (about one-fifth) of the atmosphere. There is a good reason that oxygen gas is as abundant as it is in the atmosphere but less abundant than nitrogen gas. Oxygen molecules are held together with a double bond between the two oxygen atoms. A double bond is stronger than a single bond but not as strong as a triple bond. The strength of the double bond allows oxygen to accumulate to a concentration of 21 percent of the atmosphere but not to concentrations any higher than that.

Important Point

The double bond in an oxygen molecule, depicted $:\ddot{O}=\ddot{O}:$, is strong enough to keep O_2 from being too reactive, hence unable to accumulate in the atmosphere, but weak enough to allow O_2 to be the oxidizing agent in combustion processes.

The level of 21 percent is actually an optimum value. Any less oxygen and respiration in living organisms would be much more difficult. You have probably experienced that effect if you have ever been at a high altitude. The relative concentration of oxygen gas at high altitudes is still 21 percent, but the total number of all gas molecules, including oxygen, is much less than it is at sea level, hence the difficulty breathing.

Conversely, if the concentration of oxygen in the atmosphere were much higher than 21 percent, fires would be very difficult to extinguish. Wildfires, in particular, would burn much more out of control than they do otherwise.

Is it just a coincidence that oxygen gas is present in the atmosphere at a relatively optimum level of 21 percent? Probablynot. The biosphere has a huge number of feedback mechanisms that regulate the concentration of chemical compounds found in nature. In many cases, those feedback mechanisms work toward optimizing the conditions that make Earth habitable. (If you are interested in this topic, you might read about the Gaia effect, which asserts that the activities of living organisms on Earth in fact help to create and maintain the conditions on Earth that are necessary for life to thrive.)

Argon

If you have been doing the math, you probably have noticed that 78 percent N_2 and 21 percent O_2 add up to 99 percent. There is still 1 percent of the atmosphere that is unaccounted for. Had you lived in the late eighteenth century, when nitrogen and oxygen were first discovered, you would have searched for that other 1 percent and probably would have been baffled (if not quite frustrated!) that you were unable to account for that remaining 1 percent. In fact, it was another one hundred years before the 1 percent was finally accounted for.

The identification of N_2, O_2, and other common gases like CO_2, SO_2, and NH_3, was possible because those gases undergo chemical reactions. They can be identified from their reactions and the products they form. The defining characteristic of noble gases, however, is their relative chemical nonreactivity. The noble gases escaped detection, and were not discovered until the 1890s. When the noble gases were finally discovered, it was realized that the remaining 1 percent of the atmosphere consisted mostly of argon gas, as monatomic Ar molecules, which is chemically nonreactive. (The other noble gases—He, Ne, Kr, Xe, and Rn—are also in the atmosphere, but in very tiny quantities.)

Argon's presence in the atmosphere is due to the radioactive decay of potassium-40 in Earth's crust. The process involved is called electron capture (see Chapter 13) and is shown by the following equation:

$$K + e^- \rightarrow Ar.$$

The half-life of potassium–40 is 1.28 billion years, which is about one-third the age of Earth. With such a long half-life, the amount of potassium-40 still remaining in Earth's crust is about one-eighth what would have been present in the primordial Earth. There is sufficient potassium-40 still remaining to continue to add argon-40 to the atmosphere for several billion years more.

To have achieved a relative concentration of 1 percent in the atmosphere, three things must be true of argon:

➤ Argon has to be chemically unreative, otherwise argon atoms would combine with other elements present in the atmosphere and it would eventually be removed.

➤ Argon has to be insoluble in water, otherwise argon would wash out of the atmosphere when it rains.

➤ Argon has to be too heavy to escape into outer space. On a very large scale, the atmosphere is held in place by the force of gravity. Extremely light gases that have worked their way into the upper limits of Earth's atmosphere can escape into space. For all practical purposes, hydrogen—either as individual atoms or as H_2 molecules—is the only chemical species light enough to do that. An argon atom is 40 times heavier than a hydrogen atom is, so Earth's gravity is sufficient to keep argon in Earth's atmosphere.

Historical Note

Nitrogen was discovered in 1772; oxygen was discovered in 1774. Scientists knew, however, that they had not accounted for 1 percent of the atmosphere's gases. Because argon is chemically inert, it was not until 1894 that argon was finally discovered.

Carbon Dioxide

At first look, it might appear that we have now accounted for all of the gases in the atmosphere, since 78 + 21 + 1 = 100 percent. Those numbers, however, represent the concentrations rounded off to whole numbers. If we express the percentages more exactly to several decimal places, then the sum so far is still a little less than 100 percent—but only slightly less. Again ignoring water vapor, the sum of the quantities of all the other gases in the atmosphere add up to only about 0.03 percent, and most of that is carbon dioxide (CO_2).

Taking the concentration of CO_2 to be 0.03 percent (which, again, is not an exact number, but an approximate value) might make it seem as though CO_2 is rather unimportant. In fact, the presence of CO_2 in Earth's atmosphere is much more important than the presence of Ar. After all, argon does not do anything in the atmosphere. Argon does not react with other chemical species in the atmosphere. For all practical purposes, argon does not absorb sunlight impinging on Earth's atmosphere, nor does it absorb the infrared radiation Earth is emitting back to outer space. The presence of argon has no effect on any biochemical cycles on Earth's climate, on ozone, or on any other environmental phenomenon. Carbon dioxide, however, is absolutely crucial to living organisms and to the regulation of Earth's climate.

Carbon dioxide in the atmosphere is the principal source of carbon in plants, which obtain carbon through photosynthesis. Animals obtain that carbon indirectly by eating plants or by eating other animals that have eaten plants. Animals, and plants, return carbon dioxide to the atmosphere through respiration.

Carbon dioxide dissolves in both freshwater and sea water and makes those environments just slightly acidic. As carbon dioxide dissolves, it forms bicarbonate (HCO_3^-) and carbonate (CO_3^{2-}) ions, the latter being incorporated into the shells of aquatic mollusks and other shellfish.

Along with water vapor, carbon dioxide is the principal regulator of climate on Earth. Earth gives off heat into outer space in the form of infrared radiation in order to maintain a state of thermal equilibrium. Carbon dioxide and water are both good absorbers of infrared radiation and contribute to what is called the greenhouse effect (see Chapter 29). Without those gases in the atmosphere, too much heat would be lost to outer space and Earth's surface would be too cold to support the presence of liquid water. Life might still exist on Earth, but the climate would not be at the comfortable levels to which humans have become accustomed.

Too little CO_2 (as is the case on Mars), and Earth would be too cold—water would only be present as ice, which would make it difficult for Earth's surface to support life as we generally know it. Too much CO_2 (as is the case on Venus), and Earth would be too hot—water would only be present in the gaseous state, which would also make it difficult for Earth to support life as we know it. It's the Goldilocks effect: Venus is too hot, Mars is too cold, Earth is just right. To a large extent, it is CO_2 that is helping to maintain the right balance.

Other Gases

The remaining gases that occur in the atmosphere tend to be present in very, very tiny concentrations. A concentration of 1 ppm of SO_2, for example, means that for every 1 million air molecules, there are about 780 thousand molecules are N_2, about 210 thousand molecules are O_2, about 10 thousand molecules are Ar, about 300 molecules are CO_2, and only 1 molecule of SO_2 (with any other trace gases being present at about the same level as SO_2 or less).

These remaining gases principally include oxides of nitrogen (N_2O, NO, NO_2, and the like), oxides of sulfur (SO_2 and SO_3), ozone (O_3), hydrogen sulfide (H_2S), ammonia (NH_3), methane (CH_4), several small organic molecules, the other noble gases, and just traces of various species.

There are natural and anthropogenic sources of these substances. Nitrogen-containing compounds can be formed by the action of lightning on atmospheric N_2 and by metabolic processes in living organisms. Oxides of sulfur result from geothermal activity, principally volcanism. Ammonia, methane, and hydrogen sulfide result from dead and decaying organic matter in anaerobic environments. Some atmospheric compounds are themselves formed in the atmosphere.

Anthropogenic sources of oxides of nitrogen and sulfur are principally from the combustion of fossil fuels. Much of what humans add to the atmosphere is considered to be pollution. Anthropogenic chemical compounds that have no natural sources include chlorofluorocarbons, which are chemically nonreactive in the troposphere but can damage the fragile layer of ozone in the stratosphere (see Chapter 28). Because there are no natural sources of fluorine or fluorine-containing compounds in the atmosphere, we know that all fluorine-containing compounds detected in the atmosphere must be anthropogenic in origin.

Conclusion

We may stand on the ground and swim in lakes and oceans, but our heads are in the atmosphere virtually 24 hours a day, 365 days a year. The mix of nitrogen, oxygen, and other gases is crucial to the survival of humans as well as all other forms of life on Earth.

CHAPTER 27

 # The Chemistry of Acid Rain

> ## In This Chapter
>
> ➤ CO_2 in aquatic ecosystems
>
> ➤ Acid rain characteristics
>
> ➤ Sources of H_2SO_4 and HNO_3
>
> ➤ How acid rain harms Earth
>
> ➤ How to mitigate the effects of acid rain

Natural rainfall tends to be just slightly acidic. As CO_2 dissolves in rain droplets, sleet, and snow, carbonic acid—H_2CO_3, a weak acid—forms. The term *acid rain* refers to precipitation that contains strong acids—mostly sulfuric acid (H_2SO_4) and nitric acid (HNO_3).

Important Point

The two major components of acid rain are H_2SO_4 and HNO_3.

The Role of CO_2 in Aquatic Ecosystems

CO_2 is the most important component of the atmosphere after N_2, O_2, and water vapor. (I am ignoring argon because it is chemically inert in the atmosphere.) Carbon dioxide is soluble in cloud droplets (clouds are made of water), raindrops, dew, and any terrestrial bodies of water—lakes, ponds, rivers, streams, and oceans:

$$CO_2 \text{ (g)} \leftrightarrow CO_2 \text{ (}aq\text{)}.$$

When dissolved in water, CO_2 forms carbonic acid:

$$CO_2\ (aq) + H_2O\ () \leftrightarrow H_2CO_3\ (aq).$$

As was discussed in the section on chemistry in aquatic ecosystems (Chapter 24), after CO_2 dissolves and forms carbonic acid, H_2CO_3 can react with water to form hydronium and bicarbonate ions:

$$H_2CO_3\ (aq) + H_2O\ (l) \leftrightarrow HCO_3^-\ (aq) + H_3O^+\ (aq).$$

In each of these equations, the double arrows indicate that the reactions proceed in both directions and that equilibrium is maintained among the various carbon-containing species. The important result is that even "clean" rainwater is slightly acidic due to the formation of hydronium ions, H_3O^+. Rainwater tends, therefore, to have a pH of about 5.5.

Important Point

Natural rainfall has a pH of about 5.5. The effects of acid rain usually occur when the pH is in the range 0–4.

Chem Vocab

The term *oxides of nitrogen* refers mostly to NO and NO_2 in the atmosphere, but can include N_2O, N_2O_3, N_2O_4, and N_2O_5.

The Characteristics of Acid Rain

An environmental problem brought about by the industrial revolution has been the phenomenon of acid rain. Acid rain may be defined as rain that has a pH less than 5.5, although the deleterious effects of acid rain mostly appear when the pH gets down to 4 or less. Acid rain contains nitric acid (HNO_3) and sulfuric acid (H_2SO_4), both of which have natural and anthropogenic sources. Oxides of nitrogen—principally nitric oxide (NO) and nitrogen dioxide (NO_2)—are produced by combustion of fossil fuels at high temperatures. (The Environmental Protection Agency [EPA] lumps all oxides of nitrogen together as NO_x.) Wildfires are also a source of oxides of nitrogen. Once in the atmosphere, additional reactions convert NO_2 into N_2O_5, which dissolves in water to form HNO_3, as shown in the following equation:

$$N_2O_5\ (g) + H_2O\ (g) \rightarrow 2\ HNO_3\ (aq).$$

HNO_3 droplets remain dissolved in water vapor, cloud droplets, and rain droplets.

Acid rain tends to be produced miles—sometimes hundreds of miles—downwind from centers of heavy industry or power plants. The acid rain falling in the Adirondacks, for example, comes from coal-fired electric power plants along the Ohio River Valley, several states away from the Adirondacks.

Many areas may have a thick topsoil and rocks composed of sedimentary materials, like limestone, that can neutralize or buffer the acid runoff, greatly decreasing the harmful effects.

In areas where acid rain has done the most damage, though, the landscape is characterized by a combination of thin topsoil and igneous rocks (granite, for example) that have no buffering capacity. Consequently, acid rain does its greatest damage to areas that are downwind from sources of NO_x and SO_2 and have a prevalence of soil and rock that cannot buffer the acid.

Acid deposition can be in the form of solids, liquids, or gases. If an acid is a weak acid, it only slightly lowers the pH of snow, sleet, rain, or atmospheric water vapor. If an acid is one of the strong acids—usually HNO_3 or H_2SO_4—its effect on pH is more pronounced. Any kind of acid deposition, the most familiar of which is acid rain, has both natural and anthropogenic sources.

Natural Sources of Sulfuric Acid

There are completely natural sources of sulfuric acid in the environment. Sulfur can occur in compounds on land, in water, and in the atmosphere, and is cycled through rocks, soils, both freshwater and marine systems, and the atmosphere in biogeochemical cycles.

Chem Vocab

The term *oxides of sulfur* refers to SO_2 and SO_3 in the atmosphere.

Sulfuric acid is produced naturally mostly by volcanism. Volcanic eruptions spew sulfur dioxide (SO_2) into the atmosphere. During the several days that elapse while SO_2 is being transported hundreds of miles from its original source, SO_2 is oxidized to sulfur trioxide (SO_3), as shown:

➤ $S\ (s) + O_2\ (g) \rightarrow SO_2\ (g);$

➤ $2\ SO_2\ (g) + O_2\ (g) \rightarrow 2\ SO_3\ (g).$

In turn, SO_3 dissolves in water to form H_2SO_4:

➤ $SO_3\ (g) + H_2O\ (g) \rightarrow H_2SO_4\ (aq).$

The sulfuric acid droplets remain dissolved in water vapor, cloud droplets, and rain droplets. The next time it rains, H_2SO_4 is washed from the sky, where it can land in aquatic systems, soil, or human structures.

Historical Note

In 79 CE, it was the poisonous gas released during the eruption of Mt. Vesuvius in Italy that killed the populace of Pompeii, not the flow of lava.

The second reaction is more complex than it looks. An important point to note in these reactions is that sulfur is in the +4 oxidation state in SO_2, but in the +6 oxidation state in SO_3 and H_2SO_4. Therefore, that second reaction is an oxidation-reduction reaction. Sulfur atoms are oxidized from the +4 to the +6 state, while oxygen atoms are reduced from the 0 to the –2 state.

Although it is true that some SO_2 also dissolves in water vapor to form sulfurous acid (H_2SO_3), sulfurous acid is a very weak acid, so its effect on pH is much less than the effect of H_2SO_4, which is a strong acid. The average level of H_2SO_3 might increase the acidity of rain by about half a pH unit, whereas commonly observed levels of H_2SO_4 can lower pH by several units.

Since the second reaction is an acid-base reaction, which by definition occurs with no change in the oxidation state of any of the elements participating in the reaction, the sulfur in the +4 state (as SO_2) has to have already been oxidized to the +6 state (as SO_3) before the reaction can take place.

There are a number of chemicals in the lower atmosphere that can oxidize SO_2 to SO_3; O_2 is used here as a representative species. The oxidation process is relatively slow and may require several days to take place, during which time the gases have been blown hundreds of miles from their sources. What is important is that H_2SO_4 is a strong acid, so its effect on pH can be appreciable.

During volcanic eruptions, sulfur-containing particulate matter may be spewed miles into the atmosphere and may take months or even years to eventually settle back down again. During that time, material from the volcano may travel hundreds or even thousands of miles and could encircle the globe. The particulate matter from Krakatoa's eruption formed a thin layer in the atmosphere that reflected sunlight and cooled Earth's surface for about two years following the eruption.

Natural Sources of Nitric Acid

Nitrogen undergoes a complicated cycle throughout the biosphere. Beginning with molecular nitrogen N_2, which constitutes about four-fifths of the atmosphere, nitrogen can be converted into other forms that include such species as NO, NO_2, NO_2^-, NO_3^-, N_2O, NH_3, and NH_4^+. Eventually, some nitric acid will be formed, for example, by the following set of reactions:

➤ N_2 (g) + O_2 (g) → 2 NO (g)

➤ 2 NO (g) + O_2 (g) → 2 NO_2 (g)

➤ NO (g) + NO_2 (g) + O_2 (g) → N_2O_5 (g)

➤ N_2O_5 (g) + H_2O (g) → 2 HNO_3 (*aq*)

Both NO and NO_2 are toxic. Nitric acid (NO) is odorless and colorless, but nitrogen dioxide (NO_2) has a pungent odor and reddish-brown color that is easily visible in polluted air. As in the oxidation of SO_2 to SO_3, the first and second reactions are both more complicated than shown, but they give an idea of how N_2O_5 is formed as the important precursor to HNO_3.

Notice the oxidation states of nitrogen in the various species shown in these equations. The oxidation state of nitrogen in HNO_3 is +5. The oxidation state of nitrogen in the other species are as follows:

➤ N_2: 0

➤ NO: +2

➤ NO_2: +4

➤ N_2O_5: +5

Just as with the formation of H_2SO_4, the formation of HNO_3 is an acid-base reaction, which, again by definition, must occur with no change in nitrogen's oxidation state. Therefore, by the reactions shown, nitrogen is slowly oxidized first to the +4 state in NO_2, and then to the +5 state in N_2O_5, which is soluble in water, forming HNO_3.

Since H_2SO_4 and HNO_3 are both soluble in water, the next time there is rain, snow, or sleet, the acids precipitate to Earth along with the moisture. The deleterious effects of acid deposition are caused almost completely by these two strong acids, not by weak acids. The only other common strong acid is hydrochloric acid (HCl).

Although chlorine is an abundant element in nature, it does not go through a biogeochemical cycle like sulfur and nitrogen do. There tends to be very little chlorine in the atmosphere (from any natural sources at least) as either the pure element or in compounds. On land, chlorine mostly stays locked up in rock salt and other minerals. In the ocean the splashing of salt water releases some salt into the air, which would include NaCl; but in NaCl, chlorine is present as the chloride ion (Cl^-), not as chlorine atoms. Chloride ions are not particularly reactive and are more likely just to fall back into the ocean than to combine with other substances in the atmosphere.

Anthropogenic Sources of H_2SO_4 and HNO_3

Anthropogenic sources of acid rain are the result of the combustion of fossil fuels. Atmospheric chemists usually divide the sources into two broad groups: stationary and mobile.

Historical Note

The Clean Air Act of the 1960s mandated the reduction of local levels of air pollutants from power plants and factories. In response, taller smokestacks were installed. Winds blew the pollutants away from their sources, which did help solve the problem of local pollution. Spewing the exhaust higher into the atmosphere, though, just moved the air pollutants to distant localities downwind from their sources. Overall, taller smoke stacks really were not a solution.

Important Point

Scientists tend to associate H_2SO_4 with stationary sources and HNO_3 with mobile sources.

The principal anthropogenic sources of H_2SO_4 are stationary, consisting of fossil fuel–burning electrical power plants, smelters, factories, and homes burning coal—places where fossil fuels are burned to produce electricity, run machinery, or provide space heating. Sulfur is an impurity found in fossil fuels, but mostly in coal.

When coal—which is mostly carbon—burns, sulfur burns along with it to form sulfur dioxide (SO_2). In the atmosphere, the reactions previously shown take place as winds blow SO_2, SO_3, and H_2SO_4 hundreds of miles from their sources. Like H_2SO_4 from natural sources, H_2SO_4 from anthropogenic sources washes from the atmosphere in the next rain, sleet, or snowstorm.

To a lesser extent, stationary sources also produce some HNO_3. In either case, it can be the situation that acid rain is not even a problem in the immediate vicinity of its source. Once again, winds may blow these gases hundreds of miles or more from their sources. During that time, sulfur goes through its various oxidation processes until H_2SO_4 forms and precipitates to Earth in the next rain or snowstorm.

The principal anthropogenic source of HNO_3 is the exhaust from motor vehicles that are powered by internal combustion engines—cars, trucks, motorcycles, and buses—. Normally, atmospheric nitrogen is almost chemically inert. At the high operating temperature of engines, however, nitrogen gas that is in the air-fuel mixture burns to form nitric oxide. Because sulfur is largely removed from gasoline and diesel fuel, motor vehicles are a minor source of H_2SO_4. Internal combustion engines, however, are the principal source of HNO_3 because atmospheric nitrogen burns in the engines along with the fuels.

In the atmosphere, NO goes through the same reactions that it does from natural sources to eventually end up as HNO_3. Anthropogenic sources usually contribute a much greater concentration of HNO_3 to the atmosphere than natural sources. As with oxides of sulfur, winds may blow NO and NO_2 hundreds of miles before they are converted to HNO_3 and washed from the atmosphere in the next rain, sleet, or snowstorm.

Historical Note

In the 1970s, catalytic converters were installed on motor vehicles to reduce the emissions of NO_x, CO, and unburned hydrocarbons. Catalytic converters have been extremely successful.

The Regions of North America Most Affected by Acid Rain

To a large extent the northeastern United States and southeastern Canada are the regions in North America most affected by acid rain. There are two principal reasons for this.

One reason is the heavy industry in the northeastern region of the United States and the high population density east of the Mississippi River in states like Illinois, Indiana, Ohio, Pennsylvania, New York, and New Jersey. Much of the industry derives its energy from coal-fired electrical power plants—the coal coming from mines in West Virginia, and the power plants being situated along the Ohio River as a source of cooling water for the furnaces. Eastern coal has a particularly high sulfur content. However, it is more economical to burn coal from mines close at hand than to transport coal from strip mines in the western United States, despite that western coal has a lower sulfur content.

Important Point

The effects of acid rain in the United States are most pronounced in the northeastern states, especially the Adirondack Mountains of New York and the mountains of Vermont, New Hampshire, and Maine.

The other reason the Northeast is affected by acid rain is geological. When acid rain falls to Earth, it is possible for thick topsoil containing sedimentary materials—principally limestone ($CaCO_3$)—to neutralize the sulfuric and nitric acid. Chemists would say that limestone has the ability to buffer acid runoff. In that case, the water flowing into aquatic ecosystems would have a pH closer to 7 and would do much less harm.

During the last ice ages, however, the regions of North America that are now southeastern Canada (Ontario and Quebec), New England, and the Great Lakes states were scoured by thick glaciers as they worked their way south. The glaciers scraped away the topsoil and left these regions with thin, granitic soil, which has no buffering capacity. Rain or snowfall with a pH that may be less than 4 falls to the ground and drains into lakes, rivers, and ponds, creating habitats with an unnaturally low pH. In addition, acidic runoff can leach toxic metals from the soil, causing those metals as well to drain into aquatic ecosystems. These regions suffer a form of double jeopardy—they are the main places in the United States and Canada where acid rain is likely to fall, and their soil has no defense against the acid.

In contrast, central North America—the provinces of central Canada and the plains states of the central United States—are mostly agricultural. Population density is low. There is less heavy industry. And a lower population means fewer automobiles. Millennia of occupation by large grazing animals (bison, for example) and the promotion of grasslands by Native Americans (who regularly burned the grasslands to discourage the growth of woody plants) have endowed the plains with deep topsoil that is ideal for agriculture and that can effectively buffer whatever acidic precipitation might fall in the region.

The Harmful Effects of Acid Rain

Acid rain has deleterious effects on both synthetic materials and natural ecosystems. The first can result in destruction of property, with an accompanying high price tag. The second can make aquatic habitats unsuitable for fish, amphibians, and invertebrates that normally would be present. Severely acidic lakes can become so devoid of any animal species that biologists refer to them as barren lakes.

In the 1970s, the discovery of barren lakes in Scandinavian countries; Ontario, Canada; and the Adirondack Mountains of upper New York State sounded the alarms among fishermen, recreationists, and biologists to undertake the study of acid rain and the search for solutions to the problem.

Chem Vocab

A **barren** lake is one that is devoid of any living organisms—fish, amphibians, invertebrates, even plant species. A barren lake's water tends to be crystal clear because nothing is living in it.

Lowering the pH of aquatic ecosystems severely affects breeding success of fishes and amphibians, both of which lay eggs underwater. Birth defects in these classes of vertebrates have been attributed in part to acid rain. Acid runoff leaches metals from rocks and soils and washes them into lakes and ponds, where they may be toxic to aquatic organisms.

Acid rain does more than just kill aquatic organisms; it harms lichens, ferns, the leaves of deciduous trees, and the needles of coniferous trees. Forests as a whole are weakened, making plants more susceptible to infestations of disease and insects. Acid rain leaches nutrients from soil, retarding tree growth and maturation. Croplands suffer similar ill effects, resulting in a reduction of food production.

These two gases together, HNO_3 and H_2SO_4, can lower the pH of rain to values of 4 or less, roughly the same acidity as vinegar. Acid rain damages metal and concrete structures and buildings, automobile paint, monuments, and statues made of limestone or marble, which are both forms of calcium carbonate ($CaCO_3$). As occurs any time acid and metal come into contact with each other, the metal tends to dissolve. As occurs any time acid and carbonate come together, the carbonate dissolves and forms CO_2:

$$CaCO_3 \text{ (s)} + 2\ H^+ \text{ } (aq) \rightarrow Ca^{2+} \text{ } (aq) + CO_2 \text{ (g)} + H_2O \text{ } (l).$$

Acid rain has severely damaged world heritage treasures, such as the Parthenon in Greece, which is made of marble, and many other buildings and statuary of the ancient world. The monuments to the presidents in Washington, D.C., are made of marble and have been damaged by acid rain.

Mitigating the Effects of Acid Rain or Preventing Its Formation

Sometimes airplanes have flown over lakes and dropped lime (CaO) from the air to neutralize the acid draining into the lakes. This technique works but is expensive and has to be repeated if the source of acid rain continues.

What are better solutions? For one, catalytic converters on motor vehicles have done much to remove NO_x before it leaves the vehicles' tailpipes (converting oxides of nitrogen into harmless N_2 and O_2). For another, in the case of power plants coal can be washed to remove much of the sulfur. Power plant smokestacks have been retrofitted with scrubbers that contain lime or limestone. The lime reacts with SO_2, converting it into $CaSO_4$, which in turn can be processed into H_2SO_4 for industrial use. Since sulfuric acid ranks number one as the most-produced chemical in the United States anyway, this is a good source of sulfuric acid and helps protect the environment.

Conclusion

Ultimately, one of the best solutions is for Americans to wean themselves from fossil fuels. Most of the renewable energy alternatives to fossil fuels do not contribute to acid rain or to global warming, which has become an even greater concern than acid rain. Burning fossil fuels emits CO_2 into the atmosphere, which has the greatest effect on global temperatures. Burning fossil fuels also emits SO_2 and NO_x into the atmosphere, creating acid rain. Decreasing the use of fossil fuels would go a long ways toward ameliorating both problems.

What are the alternatives? Suitable sites for hydroelectric power plants have already been developed, so hydropower is unlikely to play a larger role in the total energy economy than it does now. Nuclear power fell out of favor with Americans thirty years ago, and it is unclear whether it will make a comeback. Wind, photovoltaic energy, biomass, ethanol, tides, and fuel cells all have their advantages and disadvantages, which leads us to another solution—conservation.

Using less energy altogether helps alleviate all of our energy-related problems. Much has been accomplished in recent years to promote conservation, but there is plenty of room left for improvement. Turn the thermostat down in winter, turn it up in summer. Combine errands so fewer car trips are required. Turn off lights in rooms that are unoccupied. Replace incandescent lightbulbs with fluorescent bulbs. If everyone does his or her share, we can all make a difference.

CHAPTER 28

The Chemistry of Ozone

> ## In This Chapter
>
> ➤ Ozone in the troposphere
> ➤ Air pollution
> ➤ The Clean Air Act
> ➤ Ozone in the stratosphere
> ➤ The Montreal Protocol

Ozone (O_3) is like a weed. Dandelions are beautiful flowers in natural meadows, but homeowners prefer not to have them in their gardens. Ozone is like that. Scientists want ozone in the stratosphere, but not in the troposphere. Why in the stratosphere? The answer is that ozone's presence in the atmosphere prevents otherwise harmful solar ultraviolet rays from reaching Earth's surface.

At the minimum, the UV radiation that does reach Earth's surface causes sunburn, but it is much more deleterious than that. Overexposure to UV radiation can result in melanoma (skin cancer), cataracts in the eye, and damage to crops. Hospital personnel use UV radiation to sterilize surgical instruments. Intense UV radiation would sterilize Earth's surface. Any life form that lacked a defense against UV rays, such as being able to live in caves or underground, would die. Also, O_3 in the stratosphere does no harm because there are essentially no living organisms in the stratosphere.

Chem Vocab

Melanoma is a dangerous form of skin cancer that can result in death if not caught and treated early.

Why do we not want O_3 in the troposphere? Because that is where we live—along with all the other plants and animals on Earth! Ozone is a powerful oxidizing agent and can be the culprit lurking behind respiratory diseases, deterioration of human-made materials such as rubber and nylon, and damage to crops and other plants.

The Chemistry of Ozone in the Troposphere

Oxygen can be part of other substances in the atmosphere, but as the element itself, oxygen exists in three forms:

➤ Monatomic oxygen (O atoms)

➤ Diatomic oxygen (O_2 molecules)

➤ Triatomic oxygen (O_3)

Single-oxygen atoms are extremely transient. They are made by various reactions, but just as quickly they combine with other substances. Single-oxygen atoms, though, are necessary to form ozone (O_3). Most of the elemental oxygen in the atmosphere is in the form of O_2. Ozone is present at concentrations measured in parts per million. Wherever it exists in the troposphere, however, it tends to be harmful and is classified as an air pollutant.

Important Point

Ozone molecules that form in the troposphere tend to be very short-lived because they are so reactive. When the concentration of ozone temporarily becomes much higher than normal background levels, however, the adverse health effects on people sensitive to ozone can be rather severe. When that happens, some people have to remain indoors until the elevated ozone levels decrease again.

Pollution of the Air We Breathe

A very tiny amount of atmospheric ozone comes from natural sources. Most of it is a product of human activity. In the troposphere, ozone formation is mostly a consequence of pollution from motor vehicle exhaust. Atmospheric nitrogen (N_2) at ordinary atmospheric temperatures is essentially inert. When gasoline burns in an internal combustion engine, however, oxides of nitrogen are produced. The products are called NO_x, where x usually equals 1 or 2. That is, NO_x is a combination of NO and NO_2. The two compounds are formed as shown by the following reactions:

$$N_2 \text{ (g)} + O_2 \text{ (g)} \to 2\,NO \text{ (g), which is colorless;}$$

$$2\,NO \text{ (g)} + O_2 \text{ (g)} \to 2\,NO_2 \text{ (g), which is reddish-brown.}$$

Both gases are toxic. Other gases in the atmosphere will also oxidize NO to NO_2, so the buildup of NO_2 is fairly rapid, although remember that all of these gases are still present in the atmosphere at concentrations of only parts per million.

When visible sunlight strikes NO_2 molecules, the energy in the sunlight dissociates the molecules into NO molecules and O atoms, as shown:

$$NO_2 \text{ (g)} \to NO \text{ (g)} + O \text{ (g).}$$

Chemists call this type of reaction a photochemical reaction because light causes the reaction to occur. Oxygen atoms contain an unpaired electron (sometimes shown by a dot in the symbol, or $O\cdot$), making them very, very reactive. Chemists call species that contain one or more unpaired electrons, free radicals. Although the concentration of free radical oxygen atoms is only on the order of parts of million, the concentration of O_2 molecules is 210,000 ppm (21 percent of the atmosphere). It is true that there are four times as many N_2 molecules as O_2 molecules, but remember that N_2 is largely nonreactive at the temperature of the atmosphere.

Chem Vocab

A free radical is an atom, molecule, or ion that contains one or more unpaired electrons. Because of the tendency for electrons to pair with other electrons having opposite spin, free radicals tend to be very reactive.

Picture O atoms striking N_2 and O_2 molecules—and water vapor, CO_2, argon atoms, and dust particles. Oxygen atoms are going to react, and they are most likely to react with O_2 molecules, forming O_3 as shown:

$$O_2 \text{ (g)} + O \text{ (g)} \to O_3 \text{ (g).}$$

While these reactions are taking place, other reactions are also occurring. Other exhaust products include carbon monoxide (CO) and hydrocarbons (HC) from fragments of gasoline that were not burned completely in the engine. Again, all of these pollutants are at a very low concentration—only parts per million—but the combination results in an oxidizing atmosphere of reddish-brown haze (due to NO_2) that air pollution scientists call photochemical haze. The use of the term *photochemical* reflects the action of sunlight as a driving force behind these reactions.

Chem Vocab

A photochemical reaction is a reaction that is initiated by molecules absorbing light, which in the case of atmospheric reactions comes from the Sun.

The Daily Cycle of Air Pollution

The formation of these pollutants, their buildup in the atmosphere, and their gradual diminution follows a daily, or diurnal, cycle in urban environments that closely follows the changes in density of motor vehicle traffic during peak rush hour times. When morning rush hour traffic begins, NO, CO, and HC are emitted to the atmosphere. During the morning, NO_2 and various organic compounds, such as formaldehyde, begin to build up in concentration. As the intensity of sunlight increases during the late morning, NO_2 begins to break up into NO and O. By midday, O_3 has built up. A peak in the concentration of pollutants is reached by mid-afternoon. A shallower peak follows the evening rush, but since the Sun is beginning to set, the intensity of sunlight diminishes, so less NO, O, and O_3 is formed.

Important Point

Under breezy conditions, air pollutants are unlikely to accumulate to a level where they would pose a problem to human health. Although the concentration of pollutants from motor vehicle exhaust does increase during daytime traffic hours, their concentration tends to decrease again during the night. The biggest air pollution episodes tend to occur in very large cities with a topography that is lower than the surrounding area, allowing air to become trapped. When air is stagnant, it is easy for the level of air pollutants to rise quickly.

All day long, sunlight has been "pumping" these reactions. As soon as the Sun sets, however, the pump is turned off, and the photochemistry shuts down for the night. Many of these pollutant species, O_3 especially, are so chemically reactive that they combine with other molecules in the air, and then their concentration decreases again during the night. Breezes mix fresh air with polluted air and by morning, pollutant levels are relatively low again. As soon as rush hour traffic starts and the Sun rises, however, the cycle begins anew.

During normal weather patterns, cool air lies above warm air. Air rises during the night and breezes blow in to mix fresh air with the polluted air. Pollutant concentrations do not rise to levels high enough to cause severe health problems. If the air is stagnant, however, as occurs during a temperature inversion in which warm air lies above cool air and acts as a lid that prevents the air from rising, pollutant levels do not drop very low during the night. Therefore, the cycle begins the next morning with the pollutants already at a higher than normal level. During a temperature inversion, pollutants reach higher levels with each successive day. If this continues for several days in a row, air pollution alerts are issued. People are encouraged to stay indoors, not to drive, and to seek medical help if they begin to suffer respiratory problems.

Chem Vocab

Normally, air temperature decreases with elevation. A temperature inversion is an episode characterized by the air temperature increasing with elevation, trapping air pollutants close to the ground.

Enter the Clean Air Act

What is the solution to tropospheric air pollution? The legislation that tackled this problem was the Clean Air Act, first passed by congress in the 1960s, amended in 1970, and renewed several times since then. The Clean Air Act mandated reduced levels of the important air pollutants, among them carbon monoxide (CO), oxides of nitrogen (NO_x), and hydrocarbons (HC). To achieve compliance with the standards of the Clean Air Act, the catalytic converter was introduced in 1973 to reduce each of these three gases. (If you live in an area where smog checks are required to renew your motor vehicle registration, take a look at the certificate. You will see where the person doing the inspection checks off the levels of CO, NO_x, and HC.)

Catalytic converters have been extremely successful in reducing the level of urban air pollution. Of particular note is the reduction in the amount of CO entering the atmosphere from motor vehicle exhaust. Prior to the use of catalytic converters, asphyxiation from carbon monoxide poisoning was an ever-present concern. Poisonous carbon monoxide could enter a vehicle's passenger compartment from a leaking exhaust pipe. Anyone working on a car in a garage with the motor running and the garage door closed could succumb to carbon monoxide fumes. Couples parked on lovers' lane during the winter who left the motor running so they could run the heater sometimes died from carbon monoxide poisoning.

Historical Note

Committing suicide was sometimes done by deliberately sitting in one's car in the garage with the motor running and the garage door closed. For all practical purposes, catalytic converters have ended this. There have been recent reports of people trying to die of carbon monoxide poisoning. Their car ran out of gas, however, before they became sick at all. So catalytic converters have not only cleaned the atmosphere, they have saved lives directly.

Getting the Lead Out While We Are at It

An additional unexpected benefit of catalytic converters is that that they are incompatible with leaded gasoline. The compound tetraethyl lead, $Pb(C_2H_5)_4$, had been added to gasoline since the 1930s to improve gasoline's octane rating. It was quickly found, however, that lead coated the surface of the catalytic converter with a thin layer that poisoned the catalyst and rendered it useless. In fact, it takes only one tankful of leaded gasoline to destroy the catalytic surface. Therefore, unleaded gasoline was introduced. Subsequently, the level of lead—which is a toxic substance that causes neurological harm—in our urban atmospheres has become almost completely absent. Thus in addition to eliminating photochemical haze, a second health benefit was achieved.

Ozone in the Stratosphere

In 1930, the English geophysicist Sydney Chapman (1888-1970) worked out the cycle of ozone formation and destruction in the stratosphere. There are four chemical reactions in the cycle:

➤ Initiation: O_2 (g) + short-wavelength UV → O (g) + O (g)

➤ Formation: O (g) + O_2 (g) → O_3 (g)

➤ Destruction: O_3 (g) + long-wavelength UV → O (g) + O_2 (g)

➤ Destruction: O_3 (g) + O (g) → O_2 (g) + O_2 (g)

Short-wavelength ultraviolet radiation (light with a wavelength less than 210 nm) is the pump that drives the process. The single-oxygen atom species (O) that form are extremely reactive. They can collide either with an O_2 molecule and form ozone (as shown in the second chemical reaction in the cycle), or they can collide with an O_3 molecule and destroy ozone (as shown in the fourth chemical reaction in the cycle). Because the number of O_2 molecules is significantly greater than the number of O_3 molecules, ozone formation is favored over ozone destruction.

Important Point

Visible light has wavelengths ranging from about 400 to 700 nm. Long-wavelength ultraviolet radiation picks up where visible leaves off at wavelengths less than 400 nm. Short-wavelength ultraviolet radiation has even shorter wavelengths, less than about 210 nm. The shorter the wavelength, the higher the frequency. The higher the frequency, the higher the energy. The higher the energy, the more harmful radiation can be.

Although it is true that the absorption of ultraviolet radiation with a wavelength of about 220 to 320 nm destroys ozone molecules, it is precisely this ability of ozone that makes ozone in the stratosphere so important. Although some UV radiation still reaches Earth's surface (and causes sunburns, for example), most of the radiation is absorbed before reaching the surface, protecting living organisms on the surface from the harmful effects of UV radiation.

Catalysts That Destroy Stratospheric Ozone

The Chapman mechanism for ozone formation and destruction is incomplete. There are compounds from natural sources—N_2O from agriculture, and OH from water, for example—that contribute to ozone's destruction. In an unpolluted stratosphere, a balance between formation and destruction is maintained. Nothing can be done to promote the formation of ozone in the stratosphere. Several pollutants, however, can very effectively promote the destruction of ozone.

Important Point

Ozone can be produced very easily in the troposphere. However, it is too reactive in the troposphere for any of it to reach the stratosphere. We do not have a way to produce ozone in the stratosphere. Several chemical compounds of human origin, however, can destroy the stratosphere's ozone.

Oxides of Nitrogen

In the late 1960s, University of California physical chemist Harold S. Johnston (1920–) and Dutch atmospheric chemist Paul J. Crutzen (1933–) independently proposed that emissions of nitrogen oxides from supersonic transport aircraft (SST) flying through the stratosphere would harm the stratosphere's protective layer of ozone.

The mechanism Johnston and Crutzen proposed is the following:

➤ $NO\ (g) + O_3\ (g) \rightarrow NO_2\ (g) + O_2\ (g)$

➤ $NO_2\ (g) + O\ (g) \rightarrow NO\ (g) + O_2\ (g)$

➤ Repeat these reactions tens of thousands of times.

The effect of a fleet of SSTs would be deleterious to the stratosphere's ozone. What eventually killed the SST program, though, was economics—in the mid-1970s, fuel became too expensive.

Chlorine

In 1974, University of California physical chemist F. Sherwood Rowland (1927–2012) and his postdoctoral assistant Mario J. Molina (1943–) published a paper alerting the world to the possibility that chlorine atoms from chlorofluorocarbons (CFCs) could be destroying stratospheric ozone. CFCs, also called Freon, are compounds with formulas like CCl_2F_2 and $C_2Cl_3F_3$ that had been used since the 1930s as the working fluid in refrigeration systems and in building and motor vehicle air-conditioning systems. By the 1960s they were also commonly used as propellants in aerosol spray cans. The attraction of CFCs was that they were relatively inexpensive and they are chemically inert and nontoxic. Being inert, they did not contribute to corrosion of the appliances or AC systems in which they were used.

The chemical inertness of CFCs posed a problem, however. Refrigeration and air-conditioning systems sometimes leak. Disused refrigerators tend to end up in landfills, automobiles in junk yards. As CFCs leak into the troposphere, they gradually accumulate because they do not react with any other chemical, nor are they water soluble and therefore washed from the atmosphere when it rains. Instead, over the course of decades CFC molecules gradually drift into the stratosphere. That is where the problem with them begins. In the stratosphere, CFCs are exposed to the same ultraviolet light that O_2 and O_3 are. Ultraviolet light dissociates CFC molecules just as it does O_2 and O_3 molecules. Free chlorine atoms are liberated, and, unfortunately, right at the same elevation where ozone is most concentrated.

The mechanism Rowland and Molina proposed is completely analogous to the mechanism for NO molecules:

➤ $Cl\ (g) + O_3\ (g) \rightarrow ClO\ (g) + O_2\ (g)$

➤ $ClO\ (g) + O\ (g) \rightarrow Cl\ (g) + O_2\ (g)$

Repeat these reactions tens of thousands of times.

The discovery of the harmful effect of CFCs on stratospheric ozone resulted in the elimination of CFCs from aerosol spray cans (with the exception of essential medical uses such as asthma inhalers) and refrigeration systems. It was also discovered that similar bromine-containing compounds called halons that were used in fire extinguishers posed the same threat as Freon.

Paul Crutzen, Sherwood Rowland, and Mario Molina shared the 1995 Nobel Prize in chemistry for their work in atmospheric chemistry.

Historical Note

Harold Johnston might have shared the Nobel Prize with Crutzen, Rowland, and Molina were it not for the restriction that no more than three people can share a Nobel Prize. Also, Johnston was ineligible for his own prize later because another restriction is that a Nobel Prize cannot be given twice for the same accomplishment.

The Hole over Antarctica

In the mid-1980s, a team of British scientists working at McMurdo Research Station in Antarctica discovered that a significant decrease in the concentration of ozone over Antarctica was occurring. This was referred to as the ozone hole.

An American research team led by University of Colorado atmospheric chemist Susan Solomon (1956–) went to McMurdo in 1986 and established the link between atmospheric chlorine and the decrease in ozone concentration. The so-called smoking gun evidence of the link was the detection of chlorine monoxide (ClO) in the atmosphere over Antarctica. Since there are no natural sources of ClO in the atmosphere, it could only have been formed by the reaction shown in the first step of Rowland and Molina's proposal.

Conclusion

Enough countries have agreed to eliminate CFCs from their products that scientists fully expect the ozone layer to recover. Enough CFCs are in the atmosphere already that recovery will take several decades, but ozone levels will recover. This has been an excellent example of the application of good science to a global problem and to the problem's solution.

CHAPTER 29

Global Climate Change

In This Chapter

➤ The balance of heat on Earth

➤ An explanation of the greenhouse effect

➤ The effect of fossil fuels

➤ The effects of global climate change

For thousands of years of human history, it was assumed that the biosphere was so enormous that nothing human beings could do would alter it in any significant way. Although it was true that pollution had become an issue, pollutants from automobile exhaust, fossil fuel power plants, and factories were fairly localized problems. All that was needed was a good strong wind for a few days, and those problems would go away.

Toward the latter part of the twentieth century, however, that attitude began to change. The threat to the protective shield of ozone in Earth's stratosphere was a wake-up call to the reality that human activity could lead to the destruction of life on Earth. Recognition that Earth's oceans are not an unlimited reservoir for our garbage and sewage was another wake-up call.

During the Cold War, the threat of nuclear warfare on a global scale lead to the sobering realization that humans really do have the potential to cause the extinction of the human race, if not all life on Earth. The specter of accelerated global warming was added to the list of threats to the continuation of human existence along with many other species that are as vulnerable, or even more vulnerable, to the deleterious effects of a significantly hotter planet.

The Changing Climate

Climate has always changed. It was a much hotter world when *Tyrannosaurus Rex* walked the planet. Ice ages come in cycles. The world cools down for several centuries, the polar ice caps extend into temperate regions of the planet, and the world stays cold for hundreds, even thousands, of years. Then Earth warms up again, and the planet spends thousands of years in an interglacial period. In time, the cycle repeats itself. For the past 10,000 years, Earth has been warming up from the last ice age. There is every reason to believe that there will be more ice ages in the future.

So what is different about the current warming trend? Why have climate scientists been so much in the news for at least a decade crying that Earth is getting too hot? Why did former Vice President Al Gore receive the Nobel Prize for his efforts to educate people about global warming?

Important Point

For thousands of years humans believed that the atmosphere and the oceans are so enormous that they could not be impacted by human activity. Only in the past half century have we learned otherwise.

There are at least three important differences between past climate changes and what we are observing today. One difference is the accelerated rate at which the planet is getting warmer. In the past, warming (or cooling) trends have occurred at glacial speeds (pun intended) that have taken centuries. Life on Earth had time to adjust to the changing temperatures.

The second difference is more important to humans. All of human civilization, agriculture, population growth, and expansion of the human population into the far corners of the planet have occurred since the glaciers began to retreat from the last ice age. There are about 7 billion people who need to be fed. Global warming could severely disrupt food production. Modern humans have never experienced an episode of accelerated climate change or the effects of such a change. It is unclear whether we have sufficient time to react if we are to save ourselves.

The third difference is the most important one. We, the human race, are reputedly responsible for the global warming that is occurring. During all of the previous episodes of climate change on Earth, warming or cooling trends were due to solar activity or other natural events like volcanic eruptions or impacts of asteroids. All of these episodes were independent of any particular species of beings that lived on Earth. The notion that the activities of any one species, in this case *Homo sapiens*, can make Earth uninhabitable is completely new in the history of life on Earth.

The Natural Balance of Heat on Earth

An object in thermal equilibrium remains at a constant temperature. Although there are diurnal temperature changes at any location on Earth, and the temperature varies widely from place to place on Earth, we can treat Earth's global temperature averaged over the course of a year as being fairly constant, about 15°C (59°F). On any given day for 365 days a year, somewhere on Earth light is being received from the Sun. Light carries energy. Some of that energy is reflected back into space by clouds, snowfields, desert sand, and rocks. Some of that energy, however, heats Earth's atmosphere, oceans, and land surfaces.

Any object that is absorbing energy warms up; its temperature tends to increase. Therefore, if Earth only absorbed energy and did not radiate energy back into space, Earth's temperature would continue to increase until everything on Earth cooked. Of course, that does not happen. There is an energy balance—energy in equals energy out. From the perspective of the planet as a whole, Earth's average temperature stays relatively constant. There are annual fluctuations, and certainly some years are warmer or colder than other years. On the average, however, the average temperature is about 15°C.

Chem Vocab

Infrared is the part of the invisible spectrum that consists of longer wavelengths than the red end of the visible spectrum.

Important Point

That all objects at room temperature emit infrared radiation is the principle behind infrared photography. An infrared camera is able to record images in the infrared region of the spectrum.

Objects over a wide temperature range that includes 15°C emit heat energy as infrared radiation. I am radiating energy in the infrared as I write this. You are radiating infrared energy as you read this. Buildings, trees, rocks, plants, and animals are all radiating infrared energy.

Of the Sun's radiation that is incident on Earth's atmosphere, about 28 percent is reflected back into space and about 72 percent is emitted as infrared radiation. Earth's atmosphere plays a crucial role, however, in maintaining that constant average global temperature of 15°C. Without an atmosphere, Earth's average global temperature would be about 33°C colder—about –18°C (–0.4°F). At that temperature, the oceans still would not freeze because of their salt content. Freshwater systems, however, would remain frozen much more of the time. Agriculture would be severely affected because the growing season would be much shorter. On the plus side, I suppose, ski resorts would thrive with a substantially longer season!

Important Point

Many forms of life on Earth are adapted very precisely to the conditions of temperature and moisture under which they have evolved. Although many species can adapt to gradual temperature changes, abrupt changes could send them into extinction.

The Greenhouse Effect

Why does the atmosphere play such an important role in keeping Earth warmer than it would be without an atmosphere? To answer that question, consider the analogy of a greenhouse that is used to raise plants. A greenhouse has a glass roof. Sunlight passes through the roof and warms the interior of the greenhouse. Like everything else in the world, plants and the interior of the greenhouse radiate energy in the infrared. The glass roof, however, blocks some of the infrared radiation from escaping. The result is that the greenhouse stays much warmer than it would without a glass roof.

Earth's atmosphere functions in a manner analogous to a greenhouse's glass roof. A significant amount of the outgoing infrared radiation is absorbed by gases in Earth's atmosphere. From there, the radiation serves to either heat the atmosphere or is radiated back to Earth's surface, keeping Earth as a whole 33°C warmer than it would be without an atmosphere.

The Goldilocks Effect

Compare Earth to its nearest planets—Venus and Mars. The atmosphere of Venus is so thick that observers cannot see the surface of Venus from Earth. (Our only images of Venus's surface have come from probes that parachuted through Venus's atmosphere to the surface, and then promptly melted.) As a result, Venus experiences a runaway greenhouse effect that keeps Venus too hot to support liquid water or (probably) life as we know it. Humans could not live on Venus.

On the other hand, at the present time Mars has very little atmosphere. (It is believed by some planetary scientists that Mars may have had an atmosphere in the past but lost that atmosphere to outer space.) Consequently, there is no greenhouse effect on Mars, and Mars stays permanently cold. Mars is so cold that all surface water on Mars is frozen. Humans could live on Mars only in space suits or in artificial shelters.

Suppose the fairy-tale character Goldilocks were to visit Venus, Mars, and Earth. In the fairy tale, she said that Papa Bear's soup was too hot, Mama Bear's soup was too cold, but Baby Bear's soup was just right (and she ate all of it, much to the dismay of Baby Bear). Goldilocks would draw the same conclusions about these three planets—Venus is too hot, Mars is too cold, but Earth is just right. The whole idea behind the efforts to reverse any accelerated global warming trends is to keep Earth the way it is—just right for the life that currently inhabits it.

Greenhouse Gases

As we saw in Chapter 26, Earth's atmosphere is composed of a mixture of gases—mostly N_2, O_2, and Ar—with a variable amount of water vapor and only trace amounts of all other gases. It is the water vapor and some of the trace gases that are able to absorb infrared radiation. Because these gases are the source of Earth's greenhouse effect, they are called greenhouse gases.

The most important greenhouse gas is CO_2. Other important greenhouse gases include water, methane (CH_4), nitrous oxide (N_2O), and chlorofluorocarbons (especially $CFCl_3$ and CF_2Cl_2). Methane and nitrous oxide have natural origins, but chlorofluorocarbons are entirely anthropogenic in origin (see Chapter 28). Only about 5 percent of outgoing infrared radiation makes it through the atmosphere to outer space. Greenhouse gases absorb the remaining 95 percent of outgoing radiation, and most of that goes back into heating the atmosphere and Earth's surface.

Each greenhouse gas has a different capacity for absorbing infrared radiation and absorbs it at different infrared wavelengths. The net effect of a particular gas is the product of its infrared absorbing capability times the gas's atmospheric concentration. The concentrations of CH_4 and N_2O are significantly lower than the concentration of CO_2, but their absorbing capacities per molecule are much higher than the absorbing capacity of CO_2. We have much less control over atmospheric CH_4 or N_2O concentrations, which is why climate scientists' major focus tends to be on CO_2.

The Effect of the Combustion of Fossil Fuels

Using ice cores from Greenland and Antarctica, climate scientists can derive estimates of Earth's temperature and the concentration of CO_2 in the atmosphere going back several hundred thousand years. All the data demonstrate a convincing correlation between mean global temperature and CO_2 concentration: lower than average global temperatures correlate with lower than average CO_2 concentrations; higher than average global temperatures correlate with higher than average CO_2 concentrations.

During times of glacial advance, CO_2 concentration was relatively low. During interglacial periods, CO_2 concentration was significantly higher. CO_2 concentration was higher during the Triassic and Jurassic periods when dinosaurs ruled life on Earth.

Atmospheric scientists have documented a steady rise in the concentration of atmospheric CO_2 during the past half century, from about 315 ppm in 1958 to 390 ppm in 2012. During the same years, the rate of combustion of fossil fuels has increased proportionately to the rate of increase in CO_2 concentrations. The correlation makes sense since burning coal, oil, and natural gas is a major source of CO_2. Unfortunately, societies in developed countries have become extremely dependent on fossil fuels as their major source of energy. In the United States, about 80 percent of all energy is derived from burning fossil fuels. Efforts are underway worldwide to try to reduce fossil fuel consumption.

Important Point

Bubbles of air trapped in the ice fields of Antarctica and Greenland provide scientists with records going back hundreds of thousands of years. CO_2 concentrations can be measured directly. Temperatures can be estimated from the ratios of oxygen isotopes found in the air bubbles. The fossil records demonstrate a clear relationship between CO_2 and temperature—lots of CO_2 and high temperatures go together.

Expected Effects of Global Climate Change

There are so many variables that affect climate change—both short-term and long-term—that it is difficult to separate the influences of human activities from natural influences. However, climate scientists manage to do just that, and they attribute the changes they are seeing to the increased concentrations of CO_2, CH_4, and N_2O in the atmosphere.

One of the most noticeable changes is the melting of glaciers and polar ice. In the United States, for example, glaciers at Glacier National Park in Montana are retreating so rapidly that in a few more decades the park may be totally devoid of glaciers. Greenland's land mass is almost entirely covered with ice, which also is melting. The weight of Greenland's ice is so huge that Greenland will actually rise in elevation as the weight of the ice is removed. The extent of the ice that covers the Arctic Ocean each winter is diminishing. Before long, commercial shipping across Arctic regions should become possible.

All the water that has been locked up for hundreds of centuries in glaciers has to go somewhere, and it is. It is pouring into the oceans. With the melting of ice, sea level is expected to rise. A large percentage of Earth's population lives in coastal regions. As sea level rises, those regions will experience flooding, intrusion of salt water into freshwater supplies, disruption of agriculture, and displacement of their populations.

As Earth warms, climate changes are expected to occur. More water will evaporate from the oceans. More moisture in the atmosphere will result in more frequent violent storms—hurricanes, cyclones, and heavy monsoons. Because of disruptions to global atmospheric circulation patterns, some areas of Earth will become much wetter, while other areas will become much drier.

Middle latitudes may become too warm to support agriculture, while higher latitudes will experience a longer and warmer growing season. There will be shifts in native vegetation. As regions of Earth become warmer, plant life that cannot adapt to warmer temperatures will die out. The same thing is true of plant life that varies with elevation. Higher elevations will become warmer, so the mix of plants that can live there will change.

Animal life will also be affected. Already, migratory birds are flying north earlier each spring because of warmer weather. Birds are breeding earlier than they have historically. The main food of polar bears is seals. Polar bears are habituated to hunting seals from ice floes. Disappearing Arctic ice means fewer hunting opportunities, leading to serious concerns that polar bears could become extinct within the next century.

Animals that live at low elevations are moving up the sides of mountains where temperatures are cooler. Warmer temperatures and wetter climates also promote higher insect populations. In the cases of insects that are vectors for diseases like malaria, the incidence of insect-borne diseases can be expected to increase.

Many of these changes have already been observed. Whether or not these trends can be reversed depends upon how rapidly humans respond to the problem.

Important Point

Polar bears make great "poster boys"—they have a lot of appeal to people of all ages. There are many more lesser-known, species of animals and plants, however, that are equally in danger of extinction.

Conclusion

Recognizing the threat to Earth's climate posed by the increased level of CO_2, major efforts are taking place around the world to develop sources of energy that are alternatives to fossil fuels. During the first decade of the twenty-first century, economists had been seriously looking into expanding nuclear power since nuclear fission does not produce CO_2. However, following the nuclear accidents at Chernobyl in the Ukraine in 1986 and at Three Mile Island in the United States in 1979, public support for nuclear power was so low that there has been no growth in nuclear power in the United States since the 1980s. The nuclear accident following Japan's 9.0-magnitude earthquake and tsunami in 2011 may have the same result and put a halt to any near-term expansion of nuclear power.

Assuming no growth in nuclear power (or hydroelectric power, since suitable sites have already been developed), the two best alternative energy technologies are solar and wind, neither of which are sources of CO_2 emissions. As of 2012, major efforts are underway in the United States to develop substantial solar and wind infrastructures by 2030.

In the end, it may not be concern about global warming that weans the human race from fossil fuels. It may become a matter of economics. As the price of petroleum and natural gas increases, and the price of alternative energy sources decreases, carbon-free technologies may win out simply because they may become more affordable. In the meantime, conservation measures also have a major impact and need to be pursued as much as possible.

The potentially adverse effects of global warming are real and serious. Hopefully, as governments become more aware of the problem, they will support potential solutions. The real solution, however, lies with us, the individuals. Conservation measures that individuals adopt and sustainable energy technologies that we support will go a long way to keeping Earth "just right."

PART SIX

The Basis of Life

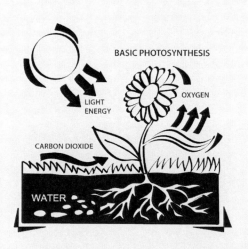

BASIC PHOTOSYNTHESIS

LIGHT ENERGY

OXYGEN

CARBON DIOXIDE

WATER

CHAPTER 30

The World of Carbon

In This Chapter

➤ Structural formulas

➤ Functional groups

➤ Hydrocarbons

➤ Compounds that contain oxygen

➤ Compounds that contain nitrogen

Organic compounds may be defined as pure substances that contain carbon and hydrogen, and perhaps other elements as well. There are millions of organic compounds. Many occur naturally, but many are synthesized in laboratories and in industrial settings. Carbon occupies a unique place in the periodic table. Carbon's position gives carbon atoms the bonding properties that permit the formation of a large diversity of compounds, making the chemistry of carbon more varied than the chemistry of any other element. It is because of the diversity and great importance of carbon compounds that a separate chapter is devoted entirely to the chemistry of carbon.

Important Point

Carbon forms more chemical compounds than any other element.

Structural Formulas of Molecules

In the examples of elements that form inorganic compounds, as discussed in the previous chapter, we usually just give a compound's molecular formula without drawing a picture of the molecule to indicate its structure. The inorganic compounds we have discussed have consisted of relatively few atoms, and a listing of those atoms—as in H_2O, $FeCl_3$, or $CuSO_4$, for example—is sufficient to identify the compound. Very large inorganic substances are not included in this book, but if they were included, knowing their structures would be important.

Molecules of organic compounds, on the other hand, often contain many more atoms than do molecules of inorganic compounds. More importantly, often there are several different ways to connect the atoms together in organic compounds—in chains or rings, for example. Each different way to connect atoms together gives a different structure to the molecule. We call these different structures isomers of a compound.

Structural formulas show how the atoms are connected together, their relative spatial orientations, and an indication of the distances between atoms and the angles between the bonds that connect atoms. Structure is the key to understanding properties of molecules, especially large molecules. Therefore, it is very common to write structural formulas for organic compounds instead of simple molecular formulas.

Chem Vocab

Isomers are compounds that have the same molecular formula, but different structural formulas. In organic chemistry, the properties of substances are determined by the structures of their molecules.

Because structure is so important, let us begin by establishing some ground rules for how structures are written. First of all, let us look at the kinds of chemical bonds that can exist between atoms. Carbon atoms can form three kinds of bonds, which are the following:

➤ Single bonds: Symbolized by a single dash, as in C – C

➤ Double bonds: Symbolized by two dashes, as in C = C

➤ Triple bonds: Symbolized by three dashes, as in C ≡ C

Since carbon atoms almost always form a total of four bonds, we can expect to see the bonding around a single carbon atom represented in one of the following ways:

$$
\begin{array}{ccc}
\mid & \diagdown \quad \diagup & \\
-\,\mathrm{C}\,- & \mathrm{C} = \mathrm{C} & -\,\mathrm{C} \equiv \\
\mid & \diagup \quad \diagdown &
\end{array}
$$

Organic compounds also contain elements like hydrogen (H), oxygen (O), nitrogen (N), phosphorus (P), sulfur (S), and the halogens (F, Cl, Br, and I). Chemists call atoms of these heavier elements heteroatoms, *hetero* meaning "different"—atoms that differ from carbon and hydrogen. Hydrogen and fluorine atoms form only one bond per atom with no exceptions. In organic compounds, chlorine, bromine, and iodine atoms most often form one bond. Oxygen and sulfur atoms most often form two bonds. Nitrogen and phosphorus atoms most often form three bonds.

Important Point

The ability of carbon atoms to form chains and rings is what gives carbon its remarkable diversity of chemical compounds.

Characteristics of Functional Groups

Although there are millions of different organic compounds, a large number of them can be easily organized around the characteristic groups of atoms that they contain, what chemists call functional groups. To try to understand the physical and chemical properties of a particular organic compound, you should start by identifying what functional group(s) it contains. Next, count how many carbon atoms are in the compound's chains or rings. Ask yourself what kind of bonding is present—single bonds, double bonds, or triple bonds. Also look for what other kinds of atoms that are attached to the carbon atoms.

For example, compounds that contain only carbon and hydrogen are called hydrocarbons. The carbon atoms can be arranged in straight chains, branched chains, or rings. To understand these arrangements, consider a carbon molecule that has six carbon atoms. There are a number of different structures that are possible. If we did not show each molecule's structure, just writing C_6H_{14}, or C_6H_{12} would not tell us which compound we were discussing. Nor would we readily see that some compounds have all single bonds, some have multiple bonds, some are branched, and some are rings.

We will continue by looking at examples of compounds with each kind of functional group. We will see how to name them, and get a feeling for some of their properties.

Counting in Greek

The prefixes for one to four carbon atoms in a chain are different than the prefixes we used in Chapter 6 for inorganic compounds. For five to ten carbon atoms, however, the prefixes are the same. The prefixes used for organic compounds are the following:

➤ One carbon atom: *meth*

➤ Two carbon atoms: *eth*

➤ Three carbon atoms: *prop*

➤ Four carbon atoms: *but*

➤ Five carbon atoms: *pent*

➤ Six carbon atoms: *hex*

➤ Seven carbon atoms: *hept*

➤ Eight carbon atoms: *oct*

➤ Nine carbon atoms: *non*

➤ Ten carbon atoms: *dec*

Examples of substances that use these prefixes include methane (CH_4) and methanol (CH_3OH), each with only one carbon atom; propane (C_3H_8) and propylene (C_3H_6), each with three carbon atoms; cyclohexane (C_6H_{12}) and 2,4-hexadiene ($CH_3 - CH = CH - CH = CH - CH_3$), each with six carbon atoms; and octane (C_8H_{18}), with eight carbon atoms.

Alkanes

Alkanes are hydrocarbons with carbon atoms that are joined together with single bonds. Their general formula is C_nH_{2n+2}. (For example, if n = 4, then 2n+2 = 10, and the formula is C_4H_{10}.) Alkanes can consist of straight or branched chains. Either way, every carbon atom has four single bonds to hydrogen atoms or other carbon atoms, and every hydrogen atom has a single bond to a carbon atom.

Important Point

Gasoline is a mixture of a large number of alkanes that have five to twelve carbon atoms per molecule. Each of those compounds also has several isomers. Octane (C_8H_{18}), for example, has eighteen isomers.

Saturated Hydrocarbons

Alkanes are called saturated hydrocarbons because they contain the maximum number of hydrogen atoms that can bond to carbon atoms in a chain. (Every carbon atom has four single bonds either to hydrogen atoms or to other atoms of carbon.) Since the chains are already, in a sense, filled with hydrogen, alkanes undergo fewer types of chemical reactions than unsaturated hydrocarbons that have double or triple bonds, hence places that more atoms can be added to the chain.

Properties of Alkanes

Some trends in physical properties can be noted. As the number of carbon atoms in the compounds increases, so does their molecular weight. As molecular weight increases, melting and boiling points increase.

At 1 atm pressure, methane, ethane, propane, and butane have boiling points below 20°C, so they are gases under normal conditions. The next heavier alkanes boil at temperatures above 20°C, so they are liquids under normal conditions. By the time the carbon chains have twenty or more atoms, the melting point is above 20°C, so those compounds are fairly solid under normal conditions.

The most important uses of alkanes are as fuels, lubricating oils, paraffin wax, and asphalt. All alkanes (actually, all organic compounds) are combustible, but the lighter ones are notably the substances found in natural gas and petroleum.

The thickness of a liquid, known as its viscosity, or resistance to flow, also increases with a higher molecular weight. The lighter liquids are suitable to be used in gasoline or jet fuels. As the substances become more viscous, they are more likely to be used as lubricating oils. The weight of motor oil is an indication of viscosity. A 10W oil has a low viscosity, and is more suitable for use in cold weather because it flows more easily. A 40W oil has a high viscosity and is more suitable for driving in hot weather.

Historical Note

In the early years of automobiles, people would drain the 10W oil from their car in the spring and replace it with a heavier oil like 30W. In the fall, they would drain the 30W oil and replace it with 10W. Multi-viscosity oils like 10–30W or 10–40W were introduced so they could be used year-round.

Another familiar property of alkanes (and of hydrocarbons in general) is that they do not mix with water, hence the saying that oil and water don't mix. Hydrocarbons tend to be less dense than water, so oil and gasoline float on water.

Cycloalkanes

Cycloalkanes are hydrocarbons that contain all single carbon-carbon bonds, but have one or more rings present. Their general formula is C_nH_{2n} (which means they have two fewer hydrogen atoms than straight- or branched-chain alkanes do). With the exception of cyclohexane, cycloalkanes are not as common as straight- or branched-chain alkanes. Cyclohexane, however, is a common laboratory solvent for substances that would be insoluble in water.

Alkenes

Alkenes are organic compounds that contain a double bond between two carbon atoms. Since its general formula is C_nH_{2n}, an alkene is an isomer of a cycloalkane that has the same number of carbon atoms. For example, n-pentene and cyclopentane are isomers.

In a sense alkenes have a shortage of hydrogen atoms and are examples of unsaturated hydrocarbons. Two important examples are ethylene and propylene. Many, many ethylene or propylene molecules can be linked together to form long chains called polymers. You have probably heard of *polyethylene, polypropylene,* or *polystyrene* (styrofoam) in connection with plastic food wraps, plastic bottles, and fast-food packages. Compared to compounds with carbon-carbon single bonds, which are relatively inert, carbon-carbon double bonds are active sites for reactions to take place. In industrial syntheses a variety of different atoms can be added to the carbon atoms that are double-bonded together, making many different useful products.

Alkynes

A common pun in chemistry is that it takes alkynes to make a world. This is true. If we can have single and double carbon-carbon bonds, we certainly can have triple bonds. Alkynes contain a triple bond, and are named using the suffix *yne*. Their general formula is C_nH_{2n-2}, so they are isomers of cycloalkenes that have the same molecular formula.

For our purposes, there is only one really important alkyne, and that is acetylene, C_2H_2, used as a fuel in welding torches. The two carbon atoms in an acetylene molecule are connected with a triple bond, as shown by the structural formula for acetylene—H–C≡C–H.

Aromatic Compounds

A diverse and important group of organic compounds is derived from benzene (C_6H_6), which has the special feature of being a six-carbon ring with three double bonds between carbons that alternate with three single bonds. Benzene differs in two important ways from cyclohexane (C_6H_{12}). For one thing, cyclohexane has only single bonds between the carbon atoms. For another, benzene is a flat, hexagonally shaped molecule—all six carbon atoms and all six hydrogen atoms lie in the same plane.

On the other hand, the hexagon in cyclohexane is puckered. If we try to define a plane in cyclohexane using any three of the carbon atoms (since it takes three points in space to define a plane), we see that the other three carbons lie either above or below the plane, and perhaps none of the hydrogen atoms lie in the plane.

Important Point

Benzene, and compounds derived from it, are among the most commonly used solvents in industry.

The Stability of Benzene

This is an especially stable structure that itself usually does not react, but which provides the backbone for a great many other compounds. Substances that contain a benzene ring are called *aromatic* compounds because of the aromas of the first compounds that were discovered. However, there are now so many known organic compounds that smell is usually of little in identifying them.

Derivatives of Benzene

If we remove one of the hydrogen atoms from a benzene ring, we are left with a phenyl group, written C_6H_5-. Other atoms, or groups of atoms, can be attached to the ring at the site of the missing carbon atom. Examples of compounds that might be formed are phenols and toluene.

Phenols

A hydroxyl group (–OH) on benzene yields the compound phenol, C_6H_5OH. The –ol suffix on phenol indicates that phenol is an alcohol. Phenol was one of the first chemicals to be used as a disinfectant. In fact, its use dates to 1865. Phenol can be pretty irritating to tissues, however, so it is no longer used for that purpose. However, other alcohols, particular rubbing alcohol (isopropyl alcohol) are still used today as disinfectants without the side effects of phenol.

A derivative of phenol is the compound urushiol. Despite urushiol's odd-sounding name, you may have had experience with it. Urushiol is the poisonous substance in poison ivy!

The compound cresol is synthesized by substituting a methyl group ($-CH_3$) for another of the hydrogen atoms on the phenol ring. Cresol also is an antiseptic, less irritating than phenol, and less expensive to manufacture. Household cleaners such as Lysol may contain cresol.

Two phenol groups linked together by two carbon atoms between them form stilbestrol, which has been used since 1939 as a synthetic female sex hormone.

Toluene

Toluene consists of a benzene ring with a methyl group substituted for one of the hydrogen atoms. The properties of toluene are similar to those of benzene. Toluene finds uses as an ingredient in paint thinner, in the manufacture of trinitrotoluene (TNT), as an octane booster, and as a coolant in nuclear reactors. Perhaps a rather unique use of toluene is to remove cocaine from coca leaves in the process of manufacturing the soft drink Coca Cola.

Alcohols

As mentioned already, alcohols are compounds that have a hydroxyl group ($-OH$) attached to a carbon atom. The most common alcohols are alkanes that have an oxygen atom inserted between one of the carbon atoms and a hydrogen atom. There are thousands, if not millions, of alcohols. However, the alcohols people encounter most frequently are methanol, or methyl alcohol (CH_3OH); ethanol, or ethyl alcohol (CH_3CH_2OH); and isopropyl alcohol (C_3H_7OH). In fact, most people just use the word alcohol for any alcoholic beverage instead of the correct name, ethyl alcohol, but everyone knows what they mean. By the way, be sure to note the continued use of the prefixes *meth*, or *methyl*, for one carbon; *eth*, or *ethyl*, for two carbons; and *prop*, or *propyl*, for three carbons.

Names like methyl alcohol, ethyl alcohol, and isopropyl alcohol are what chemists call common names. Systematic names are derived by dropping the *e* from the name of an alkane and substituting the suffix *-ol*.

Methanol

An oxygen atom makes all the difference in the world between an alkane and an alcohol. At 1 atmosphere pressure, methane (CH_4) boils at -161°C, is a gas, and is insoluble in water. Methanol (CH_3OH), on the other hand, boils at 65°C, is a liquid under normal conditions, and is completely miscible with water. In fact, methanol and water mix so completely that just looking at a solution of the two, you would not know that different substances were present. Methane and methanol are both flammable, but methane possesses a much higher heat content and gives off more heat when it burns.

Methanol sometimes is called wood alcohol because an early way of making it was to heat wood under anaerobic conditions. Methanol is used in industry as a starting material for the syntheses of more complicated substances.

One very important property of methanol needs to be mentioned; methanol is toxic. (Actually, all alcohols are toxic. Ethanol, used in alcoholic beverages, is merely the least toxic alcohol.) Methanol attacks the optic nerves and drinking it can lead to blindness. (This property of methanol was even used in an episode of the former television series *M*A*S*H*. One of the soldiers went blind, and Hawkeye Pierce traced the cause of the soldier's blindness to his having drunk bootleg methanol instead of ethanol.)

Chem Vocab

Anerobic means the absence of oxygen. Conditions that are anaerobic or conditions in which air is excluded.

Ethanol

Alcoholic beverages have been around for thousands of years and are the end product of the fermentation of fruit or fruit juices. People's first encounters with alcohol were probably by accident. Once they learned how to control the fermentation process, however, the production of alcoholic beverages was inevitable. The chemical reaction in fermentation is

$$C_6H_{12}O_6 \rightarrow 2\ C_2H_5OH + 2\ CO_2\ (g)$$

where the reactant is the sugar glucose, and the main product is ethyl alcohol.

The fermentation process itself is capable of producing beverages that are at most 15 percent alcohol. Remember that alcohols are toxic—if the concentration of alcohol exceeds 15 percent, the alcohol will kill the organisms responsible for fermentation. Ethanol boils at 78°C (172˚F), which is less than the boiling point of water (100ºC, or 212˚F). Therefore, beverages with a higher alcohol content can be produced by distilling the alcohol from fermented drinks, collecting the alcohol, and then using the alcohol to make beverages with higher alcohol content.

Alcoholic beverages are labeled according to their proof. Proof is twice the percent alcohol in the beverage. Thus, 100-proof alcohol is 50 percent alcohol. Distillation turns beer into whiskey and wine into brandy.

Grains such as corn and barley contain starch and also ferment. Because of the production from barley of beer and ale early in human history, ethanol is often called grain alcohol.

Ethanol can be purchased at hardware stores and home improvement centers under the name denatured alcohol. Denatured alcohol is 95 percent ethanol and 5 percent toxic substances that have been added to render it foul tasting and even more poisonous. If alcohol sold in this fashion were not denatured, stores would need a liquor license to sell it—and probably could not keep it on their shelves at $15 a gallon!

Like other alcohols, ethanol is an antiseptic. And, like any organic compound, ethanol is flammable and is increasingly being used as a renewable fuel. Ethanol also can be used as an octane booster and as antifreeze.

Isopropyl Alcohol

There are two isomers of C_3H_8O—n-propyl alcohol and isopropyl alcohol. Isopropyl alcohol commonly is sold as rubbing alcohol and consists of a solution that is 70 percent alcohol and 30 percent water. Rubbing alcohol gets its name because it can be rubbed on sore muscles to relieve pain. Rubbing alcohol also is a good antiseptic and is often used by doctors and nurses to disinfect a patient's skin prior to injection with a hypodermic needle.

Ethers

An ether is an isomer of any alcohol that has the same number of carbon atoms, because both kinds of compounds have the general molecular formula $C_nH_{2n+2}O$. The difference is that in alcohols, the oxygen atom is inserted between a carbon atom and a hydrogen atom, whereas, in ethers, the oxygen atom is inserted between two carbon atoms. That may seem like a minor difference, but it causes important differences to exist in the properties of the two kinds of compounds.

In general, ethers are much more volatile than alcohols are. *Volatile* substances evaporate easily. Because ethers produce more vapors, and vapor ignites more easily than liquids do, ethers are more flammable—even explosive—than alcohols are.

Ethers are much less soluble in water than alcohols are. Drinking alcohol may make a person drowsy, but breathing ether fumes will put a person to sleep completely. In fact, one of the first uses of diethyl ether (usually just called ether) was as an anesthetic to put patients to sleep during surgery.

Compounds Containing a Carbonyl Group

A carbonyl group has a carbon atom and an oxygen atom connected with a double bond (i.e., C=O). Four kinds of organic compounds that contain only C, H, and O have a carbonyl group: aldehydes, ketones, carboxylic acids, esters.

Aldehydes and ketones are very similar to each other. The difference is that, in an aldehyde, the carbonyl group is at the end of an alkyl chain, whereas in a ketone, the carbonyl group is not at the end of the chain. The best known aldehyde is formaldehyde ($H_2C=O$).

Formaldehyde is a gas that is very irritating to the eyes, nose, and throat. It is an example of a lachrymator, which is a substance that causes eyes to tear. Formaldehyde is also lethal to microorganisms, which is why formaldehyde—in the form of an aqueous solution called formalin—has been used to preserve dead organisms ranging from laboratory specimens to human cadavers.

The most common ketone is acetone (CH_3–CO–CH_3). As a solvent, acetone mixes readily with water and in many other organic substances. Acetone often is sold by the gallon in hardware stores and home improvement centers to be used as paint thinner.

Carboxylic Acids

In carboxylic acids, the combination of C=O plus an –OH on the carbon atom gives a carboxyl group, often represented –COOH. The O–H bond in the carboxyl group is weaker than the O–H bond in either C–H (as in a hydrocarbon) or O–H (as in an alcohol), so solutions compounds containing a carboxyl-group are acidic, hence the name carboxylic acid. The name of a carboxylic acid ends in the suffix *ic* (or *oic*) and the word *acid*.

The strongest carboxylic acid is formic acid, H–COOH. Formic acid is released when red ants bite and when nettles are touched, and is the source of the strong irritation people feel. In fact, formic acid is so closely associated with ants that the Latin word for *ant* is the origin of the word *formic*.

Acetic acid (CH_3COOH) was the first acid known to humans and is the acid found in vinegar. Acidic solutions are sour to the taste. The word *acid* derives from the Latin word for "sour," and the word *acetic* derives from the Latin word for "vinegar." The ethyl alcohol in alcoholic beverages like wine can turn to acetic acid, making the wine taste sour. In the days before the chemistry of winemaking was understood, sour wine was fairly common.

Historical Note

Sour wine may be the source of the story recorded in the gospels about the wine steward running out of wine at a wedding feast in the town of Cana. When Jesus of Nazareth, who was a guest at the wedding, was persuaded by his mother to turn water into wine, people marveled that the new (fresh) wine was superior to the old they had been drinking. In the wine they had been drinking, it is possible that the ethanol had turned into acetic acid, imparting a sour taste to the wine.

Carboxylic acids may contain two –COOH groups, in which case they are called dicarboxylic acids. The most familiar example is oxalic acid (HOOC–COOH). If the two hydrogen atoms are removed, the species formed is called the oxalate ion ($^-$OOC–COO$^-$). The oxalate ion forms an insoluble white precipitate with Ca^{2+}, giving calcium oxalate. Oxalate ions form as a natural part of human metabolism.

Ordinarily, small amounts of calcium ions and oxalate ions are excreted in a person's urine, along with tiny crystals of calcium oxalate that may form. In exceptional cases, however, the crystals can lump together to form kidney stones. Kidney stones can block the tubes leading from the kidneys to the urethra. Passing a kidney stone may be extremely painful, and, in some cases, surgery may be required.

Esters

Esters are formed when an alcohol and a carboxylic acid combine. For example, the following equation shows the reaction between methyl alcohol and acetic acid:

$$CH_3OH + HOOC–CH_3 \rightarrow CH_3COO-CH_3 + H_2O.$$

Notice that a water molecule is formed from the hydrogen atom on the acid and the –OH group on the alcohol. This reaction is called esterification, and is just one example of what chemists call dehydration synthesis, where the word *dehydration* means to "remove water," or a condensation reaction, because water "condenses" as the reaction proceeds.

Esters are known for their pleasant fragrances. Ethyl acetate, for example, is one of the choices of solvents used in fingernail polish remover. Acetone could be used, but ethyl acetate smells better and is easier on the skin. The pleasing fragrances in fruits and flowers typically are due to esters. Chemists often use these fragrances in manufacturing perfumes.

Important Point

Amines are often known by their foul odors, as should be apparent by compounds with names like putrescine, and cadaverine!

Compounds that Contain Nitrogen

In addition to the usual C, H, and O, amines contain nitrogen (N). Amines are derived from ammonia (NH_3) by substituting an alkyl group for one or more of the hydrogen atoms on ammonia. To name an amine, the alkyl groups are named, followed by the word *amine*. Examples include methyl amine, dimethyl amine, methyl ethyl amine, and trimethyl amine. Like ammonia, amines are weak bases.

Conclusion

The element carbon forms a greater number and a greater diversity of chemical compounds than any other element in the periodic table. Life as we know it could not exist without carbon. Earth truly is a "world of carbon."

CHAPTER 31

 # The Chemistry of Life

In This Chapter

➤ Carbohydrates

➤ Lipids

➤ Proteins

➤ Nucleic acids

Chemistry that takes place in the cells and tissues of living organisms is complex and involves very large molecules called biological molecules, or biomolecules. Biomolecules support the functions necessary for an organism to be alive. These functions include the ability to obtain energy and nutrients from the environment surrounding an organism, develop structure and grow, reproduce and transmit genetic information to offspring, and, in the case of animals at least, move along the ground or through air or water. Living organisms are composed of tissues, which, in turn, are composed of cells. At the most fundamental level, cellular function depends upon the chemical reactions that take place in cells.

Living organisms require a variety of essential inorganic substances, the most important substance being water in liquid form. Plants produce molecular oxygen, without which animals could not exist. A large number of inorganic ions are essential to living organisms. Ions that are macronutrients (required in a relatively large quantity) include Na^+, K^+, Fe^{2+}, Ca^{2+}, Cl^-, HPO_4^{2-}, and HCO_3^-. Micronutrients (required in a relatively small quantity) include Cu^{2+}, Se^{2-}, Mg^{2+}, I^-, Zn^{2+}, and Co^{2+}.

We can classify the biomolecules into four groups:

➤ Carbohydrates

➤ Lipids

➤ Proteins

➤ Nucleic acids

In this chapter we will learn about each of these groups.

Carbohydrates aka Saccharides

The term *carbohydrate* literally means "carbon and water." Carbohydrates consist of only three elements: carbon, hydrogen, and oxygen. Another name for carbohydrates is saccharides. A general formula is $C_n(H_2O)_n$, meaning that a carbohydrate molecule essentially contains n carbon atoms and the equivalent of n water molecules.

Carbohydrates have several essential functions in living organisms. One of the most important is the role of carbohydrates in the metabolic process in which sugar molecules are "burned" in the cells to produce energy. Carbohydrates are the building blocks used by living organisms to synthesize the other biomolecules. Cellulose forms the structural material that makes up the cell walls of plants. Two specific sugars—ribose and deoxyribose—are found in nucleic acids.

Chem Vocab

The metabolic process is an organic process that is necessary for life.

There are three classifications of carbohydrates:

➤ Sugars

➤ Starches

➤ Cellulose

Starch and cellulose molecules are made by linking sugar molecules together into relatively long chains. In this sense starches and cellulose are examples of biological polymers.

Sugars

When most people hear the word *sugar*, they think of table sugar, which has the chemical name sucrose. Sucrose, however, is just one of a large number of sugars. Generally, sugars are classified according to the number of carbon atoms in their chains or rings, and whether they are a simple sugar or two sugars linked together.

Simple sugars, also known as monosaccharides, contain three to six carbon atoms and are grouped as trioses, tetroses, pentoses, and hexoses. In naming simple sugars, the suffix *ose* is used. Perhaps the most familiar simple sugars are glucose and fructose, both of which have six carbon atoms. Unlike much larger carbohydrates, simple sugars are soluble in water.

Another name for glucose is dextrose, and it is the sugar used by the body to produce energy. In hospitals, solutions of glucose may be administered to patients intravenously to provide nourishment. A combination of two sugars yields a disaccharide. Table sugar (sucrose) is a familiar example. It is a combination of glucose and fructose. Sucrose is also the sugar in most fruits and vegetables. Lactose, found in dairy products, is a disaccharide made of glucose and another simple sugar, galactose.

Starch and Cellulose

Starches are polysaccharides, meaning that they are large molecules formed by linking together many simple sugars. Being large molecules, they tend to be insoluble in water and unable to be absorbed by our bodies directly. Only simple sugars are small enough to be absorbed through the walls of an animal's intestine.

Polysaccharides must be broken down during digestion into monosaccharides. To accomplish this requires digestive enzymes. An enzyme is a type of protein molecule that acts as a catalyst. Enzymes are very specific in their activity—each disaccharide requires a specific enzyme to break it down. These kinds of digestive enzymes are named using an *ase* suffix. The names lactase and sucrase indicate that those enzymes are the ones that break down lactose and sucrose, respectively.

Starches are long chains of glucose molecules that have been strung together. Plants store glucose in the form of starch. In animals, the long chains of glucose molecules are referred to as glycogen. When an animal eats a plant, the starch in the plant is broken down into simple glucose molecules, which are absorbed through the walls of the small intestine. Depending on the level of glucose in the animal's cells, some glucose is transported to the cells immediately to support metabolic processes. Any glucose that is not needed immediately is stored in the liver as glycogen. Later, as the level of sugar in the blood decreases, glycogen is broken back down into glucose molecules, which can then be transported to the cells.

Important Point

Every chemical reaction that takes place in our bodies is catalyzed by an enzyme specific to that reaction. Without enzymes, living organisms could not survive.

Cellulose is the most abundant organic compound on Earth. It forms the structural components of plants and woody tissues. An interesting fact of nature relates starch and cellulose. Both are made of glucose molecules, but the molecules are linked together in such different ways that starch and cellulose are completely different from each other. For one thing, most animals can utilize starch in their diets, but few can utilize cellulose. Even termites—which are known for eating wood—can only metabolize cellulose because of microorganisms that live in their guts.

Photosynthesis

Where do carbohydrates come from? The answer is one of the miracles of life. If plants lacked the ability to combine carbon dioxide and water in the presence of sunlight to make glucose—the process called photosynthesis—neither plants nor animals could exist. How plants "learned" the process of photosynthesis is one of science's unsolved mysteries, but without it, probably the only life on Earth would be one-celled organisms like bacteria.

The following chemical reaction is an example of what takes place in photosynthesis:

$$6\ CO_2\ (g) + 6\ H_2O\ (l) \rightarrow C_6H_{12}O_6\ (s) + 6\ O_2\ (g).$$

Not only does photosynthesis result in the production of glucose and other simple sugars, but photosynthesis also is the source of Earth's oxygen gas. The word *photosynthesis* means "synthesis in the presence of light." Both sunlight and the chlorophyll in green plants are necessary for photosynthesis to occur. The essential inorganic ion present in chlorophyll is magnesium (Mg^{2+}).

Important Point

With only rare exceptions, photosynthesis is the source of all food on Earth.

Lipids

There are several classifications of lipids, including fats, oils, steroids, and fat-soluble vitamins. Lipids are never soluble in water. Like carbohydrates, fats also are a source of energy for living organisms. In fact, fats are a more concentrated form of energy. The energy content of sugar is about 4 kcal per gram. The energy content of fat is twice that much, about 8 kcal per gram. The reason that fats are a more concentrated form of energy is that fats are almost entirely composed of just carbon and hydrogen atoms. Only a very few oxygen atoms are present (remember that carbohydrates have roughly an equal number of carbon and oxygen atoms). Since more energy is released when a C–H bond is broken than when a C–O bond is broken, fats have a higher energy content.

Lipids have other important functions:

➤ They are major components of cell walls

➤ They include the sex hormones

➤ They include fat-soluble vitamins (A, D, E, and K)

Fatty Acids

Fatty acids are made from carboxylic acids and long chains of carbon atoms, having anywhere from twelve to twenty-four carbon atoms in a chain. If the carbon chains contain only single bonds, they are said to be saturated fats. If the carbon chains contain one or more double bonds, they are said to be unsaturated fats.

Labels on food containers often specify the number of grams of total fat per serving, and then break that down in grams of saturated fats and grams of unsaturated fats. In addition, if the molecules of unsaturated fat contain only one double bond, they are labeled monounsaturated fat, and if they contain more than one double bond, they are labeled polyunsaturated fat.

Important Point

A nutritionist's calorie is actually a chemist's kilocalorie. The labels of packaged and canned foods give numbers in calories, but they are actually kilocalories.

Saturated fats tend to be solids or semisolids at room temperature and are derived from animal products. Lard is an example. Unsaturated fats tend to be in more of a liquid state at room temperature and are mostly derived from plant products. Cooking oil is an example. A minimum intake of fats in the diet is absolutely essential to human health. Nutritionists, however, generally agree that unsaturated fats are healthier for your heart than saturated fats.

Important Point

There are "good" fats and "bad" fats. Good fats tend to be derived from vegetable sources and contain unsaturated molecules. Bad fats tend to be derived from animal sources and mostly contain saturated molecules.

Waxes are derived from fatty acids. Waxes found in nature include the protective coatings on some fruits and vegetables, the thick surface of cacti and other succulents, the secretions some birds and mammals use to waterproof feathers and fur, and the beeswax in honeycombs. Derived from plants, waxes are used in polishing compounds for cars, furniture, and floors. Waxes are also used in cosmetics.

Steroids

Steroids are molecules that contain four fused rings—three rings with six carbons in them, and one ring with five carbons. As a class of compounds, steroids include cholesterol and certain hormones.

Cholesterol is a compound found mostly in animals. Despite the bad reputation cholesterol gets as a potential cause of heart disease, cholesterol actually is an essential component of cell membranes. In addition, the body synthesizes sex hormones from cholesterol. Even if a person attempted to exclude all cholesterol from his or her diet, the body would continue to manufacture all the cholesterol the body requires.

Sex hormones can be classified into two groups: male hormones and female hormones. Male sex hormones are called androgens, with testosterone probably being the most familiar. Female sex hormones include estrogens and progestins. Note, however, that males and

Important Point

Cholesterol is an essential substance for human health. Just as with fats, though, there is "good" cholesterol and "bad" cholesterol.

females produce at least tiny quantities of both male and female sex hormones.

Sex hormones control the development of the sex organs in both men and women. In men, male sex hormones control the production of sperm as well as men's deeper voices, facial hair, and muscle development. In women, female sex hormones control the production of ova as well as women's higher voices, lack of facial hair, and breast development.

You may be aware of the controversy surrounding athletes consuming a large quantity of synthetic steroids so as to enhance the production of proteins and, in turn, muscle growth. Taking a large quantity of these steroids can lead to severe side effects in both men and women.

Proteins

One important difference between proteins and the carbohydrates and lipids we have discussed so far is that all protein molecules contain nitrogen (and possibly sulfur) in addition to carbon, hydrogen, and oxygen. Proteins are the real workhorses of living organisms. Important groups of protein molecules include the following:

> ➤ Catalysts, which are called enzymes, for all of the various chemical reactions taking place inside a living organism

> ➤ Hemoglobin, the protein molecule in animals that transports oxygen through the body

> ➤ The tissues of which muscles are made

> ➤ The major components of bones, cartilage, hair, fingernails, feathers, and horns

> ➤ Protein hormones, including perhaps the best-known example, insulin, which controls levels of blood sugar in the body

> ➤ Antibodies and blood-clotting proteins

> ➤ Toxins and venoms

> ➤ The albumin in egg whites

Amino Acids

Protein molecules are composed of long chains of amino acids strung together in a highly specific order. One end of an amino acid has a nitrogen (amino) group, usually shown as $-NH_2$. The other end has a $-COOH$ group, which you may recall designates a carboxylic acid. In structural representations of amino acids, the letter *R* stands for "residue," or grouping of atoms. There are twenty amino acids found in nature, each with a different residue.

Important Point

Proteins are the workhorses of living organisms. Biological chemistry is all about proteins.

When we eat foods that contain protein, digestive enzymes break the protein molecules apart into their component amino acid molecules. One of two things can happen to the amino acid molecules after they are absorbed into the body through the walls of the small intestine. Either they can be used to synthesize other amino acids that the body needs but which were not in the original protein molecules, or, they can remain unchanged. Either way, a reservoir of amino acids is present in the body's cells that can be used to produce new protein molecules.

Essential Amino Acids

Essential amino acids are those that the human body cannot synthesize, so they have to come from foods that we eat. A diet that includes any animal products provides all of the essential amino acids. A strictly vegetarian diet, however, may lack some of the essential amino acids, because no one plant contains all of the essential amino acids. Depending on a person's diet, amino acid dietary supplements may be necessary to supply those substances. Fortunately, it is not too difficult for a vegetarian to obtain a complete mix of amino acids—any combination of rice and beans, or grains and legumes, will supply all of the essential amino acids.

How Protein Molecules Are Formed

To form protein molecules, the amino ends of amino acid molecules link together with the acid ends of other molecules. The reaction is another example of a dehydration synthesis, similar to what we have already encountered when sugars link together to form starch and cellulose. The bond that forms is called a peptide bond. Each different kind of protein molecule is characterized by the particular amino acids of which it is composed as well as the sequence in which they appear.

Important Point

Our bodies are capable of manufacturing many of the amino acids we need from the protein in the food we eat. Certain substances called essential amino acids, however, cannot be manufactured by our bodies and must be present in the food we eat to prevent protein deficiencies.

But how does an organism "know" what amino acids to use to synthesize a particular protein like insulin or hemoglobin, and what the necessary order of amino acids must be? The answer lies in the organism's DNA, located in the nuclei of cells. DNA molecules are the genetic code that stores the information necessary for protein synthesis.

Enzymes

Building up protein molecules requires enzymes. For each pair of amino acids that can link together, there is a specific enzyme necessary to catalyze the reaction. In digestion, the opposite process occurs. The amino acids in a pair are separated from each other, again through a reaction catalyzed by an enzyme specific for that particular pair of amino acids. You can begin to see why enzymes are so important!

Nucleic Acids

As the term *nucleic acid* suggests, a nucleic acid is a compound found in the nuclei of the cells of living organisms. There are two kinds of nucleic acids—deoxyribonucleic acid (DNA) and ribonucleic acid(RNA) Like carbohydrates and proteins, DNA and RNA are long-chain polymers.

DNA and RNA contain three fundamental components:

> ➤ A sugar

> ➤ A nitrogenous base

> ➤ A phosphate group

Important Point

Hereditary information, our genetic code, is stored in our DNA, which serves as a template for protein synthesis.

The two parallel strands of DNA consist of deoxyribose sugar and phosphate groups. They form the backbone of the DNA molecule and look like the long structural parts of a ladder. The nitrogenous bases link the two parallel strands together like the rungs of a ladder. The sequence of phosphate group, a sugar group, and a base makes a unit called a nucleotide. The sequence of nucleotides makes DNA. The bases are the important part because they form the genetic code stored in DNA and used as a template for protein synthesis.

Nitrogenous Bases

DNA contains four bases:

> ➤ Cytosine (C)

> ➤ Thymine (T)

> ➤ Adenine (A)

> ➤ Guanine (G)

Thus, the genetic code stored in DNA is an alphabet with four letters: C, T, A, and G. Any combination of three letters—CCT, TAG, CGG, etc.—is a "word" of code that stands for an amino acid. Sections of DNA may contain redundant codes. If we think of the sequence of words like a sentence, a sequence of three letters could function like a period in a sentence. Also, there are sections of code that appear not to have any meaning, or at least their meaning has not yet been decoded.

Important Point

An amazing fact is that if a DNA molecule were stretched out in a straight line, it would be 9 feet long!

Important Point

The speed at which enzymes catalyze the necessary reactions is incredible. An enzyme can catalyze a chemical reaction inside a living organism at many, many times the speed with which any industrial catalyst can catalyze a chemical reaction.

Altogether, DNA is a marvelous storage unit of genetic information. Cells are microscopic, and their nuclei are even smaller. Yet long, long chains of DNA molecules are coiled up inside the nuclei. An amazing fact is that every cell in a living organism contains exactly the same DNA molecules, but in each tissue or organ, parts of the DNA molecules are "turned off," and only certain parts actually are used to make protein molecules. If this were not true, then internal organs would be growing hair, skin, and muscle. All sorts of protein syntheses would be taking place in parts of the body where they do not belong, which would probably overtax any organism and lead to its premature death. So, each cell of a living organism knows just which proteins it should be making, and which proteins it should not be making.

The Role of RNA

DNA, which contains the information necessary to synthesize proteins, is found only in the nucleus of a cell. Protein synthesis, however, takes place in the cytoplasm that surrounds the nucleus. So how does the coded information get from inside the nucleus to the outside? The answer lies in the role of RNA.

There are different kinds of RNA molecules, each with a different function. Basically, though, one kind of RNA obtains the information from the DNA in the nucleus. The RNA molecule has the same sequence of bases as the DNA molecule except that RNA has uracil instead of thymine.

RNA passes through the membrane that surrounds the nucleus and enters the cytoplasm. There, the different kinds of RNA work together to bring amino acids already in the cytoplasm into the correct sequence for a particular protein molecule to be synthesized. The whole process is amazingly fast (it has to be, or else the organism would probably die). Again, enzymes are at work. Tens of thousands of protein molecules are synthesized every second.

The Question of the Origin of Life

The discussion we have just completed leaves us with a puzzling dilemma regarding the origin of life on Earth. We have seen that DNA is essential to code for the formation of protein molecules—no DNA, no proteins. However, we have also seen that a least one kind of protein—enzymes—are essential for DNA and RNA to do their work. Enzymes are necessary to synthesize DNA and RNA molecules in the first place—so no enzymes (which are proteins), no DNA.

The problem is like the age-old chicken-and-the-egg problem: which came first—the chicken or the egg? In this case, which came first—DNA or proteins? It is a problem for which science currently does not have a satisfactory answer.

One proposal is that RNA molecules formed first. Somehow, the necessary inorganic precursors at random managed to produce the first RNA molecules, and these RNA molecules were self-replicating. It is an intriguing theory. It is not my purpose here to affirm or rebut it, but simply to present it and state that it has not yet been proven.

Another problem is that scientists are not certain that conditions on Earth were ever exactly suitable for the chemical evolution that had to occur before there was any biology. In other words, we usually assume that the first living organisms were single-celled creatures like bacteria. But before there could be any single-celled organisms, all of the organic molecules essential to that first simple organism had to be self-assembled somehow from the mix of inorganic materials that were available on the primordial Earth. It is probable that a few amino acids could have formed spontaneously, but that is a very long way from a self-replicating, fully alive one-celled organism.

In fact, it is also a very long way from the first one-celled organism to the first multicelled organism. Biologists consider the appearance of the first multicellular life to be as big a step in the history of life on Earth as was the appearance of the first unicellular life.

Currently, there is considerable thinking among biologists that if conditions on Earth were never right for life to spontaneously arise, then the first living organisms had to have been "seeded" from outer space. This theory is called panspermia, and means that life presumably started somewhere else (not even necessarily in our own solar system) and then spread through space. Again, I present the theory of panspermia neither to defend nor to deny it. However, there is no proof of its being true.

You may be wondering about evolution. What about Darwin? What about the origin of species? Charles Darwin (1809–1882) wrote about the origin (evolution) of new species from preexisting species (not nonliving inorganic matter) through a process he called natural selection.

Darwin did not conjecture about the origin of life itself, or at least he did not say anything beyond simply stating that somehow life did get started on the primordial Earth. All of Darwin's examples of natural selection used relatively complex multicellular organisms—plants, birds, mammals, reptiles, etc.—to explain how over a very long period of time, species of plants and animals become extinct and are replaced by new species. Evolution by means of natural selection should not be used to explain the origin of life, but it does explain the origin of new species.

Important Point

Scientists still do not have a satisfactory explanation for how the first self-replicating living organisms appeared on Earth. Origin-of-life theories continue to be an area of active scientific research.

Darwin's accomplishments actually were quite remarkable, considering that he knew nothing of the principles of genetics, or of DNA. Consequently—and Darwin himself admitted this in his book *The Origin of Species*—Darwin was unable to explain how natural selection operates at the level of transmitting genetic inheritance from one generation to the next. There is no doubt that Darwin would be truly amazed if he could come back today and find out what biology has accomplished since his death!

Conclusion

Living organisms are amazingly complex. A myriad of chemical reactions take place every second in every organism on Earth, from simple one-celled bacteria to chimpanzees, elephants, redwood trees, and human beings. Much of human health revolves around maintaining a proper balance of the chemicals found in our bodies. The more a person understands about the principles of chemistry as applied to biochemical processes, the more he or she will be able to optimize his or her own health.

CHAPTER 32

What's On Your Dining Table?

In This Chapter

➤ Biotechnology

➤ Agricultural chemicals

➤ Genetic engineering

➤ Genetically modified organisms

➤ Recombinant DNA technology

Historically, chemists have mostly worked with minerals and other nonliving materials. Biotechnology allows the development and commercialization of chemical products derived from biological materials. Biotechnology has a very long history, in fact, a history as long as that of human civilization itself.

Genetic engineering has generated considerable controversy, not just in the United States and Canada, but around the world. The goal of this chapter is for you to become sufficiently knowledgeable about the pros and cons of genetically modified foods so that you can understand the issues and draw your own conclusions about whether or not you want genetically modified organisms (GMOs) in the food you eat.

Historical Background

The earliest biotechnical operation was the process of fermentation, in which organic compounds in foodstuffs are converted into other desired items. Cheese, yogurt, vinegar, and alcoholic beverages are all produced by fermentation. Since fermentation processes were developed before the advent of modern chemistry, they were discovered accidentally and improved upon by trial and error without any real understanding of the necessary biochemical reactions taking place.

Biotechnology removes much of the trial and error—desired products can be genetically engineered. During the past century biotechnology has focused to a large extent on two important areas: agricultural chemicals and pharmaceuticals. In this chapter you will learn about developments in agriculture. In the next chapter you will learn about the work being done by the pharmaceutical industry.

Genetic manipulation is not something new. Selective breeding of animals has been done since the dawn of early civilizations as evidenced by the variety of breeds of dogs and cats available today to pet owners. All farm animals are the descendants of wild ancestors that have been bred selectively for desired traits.

Chem Vocab

Biotechnology is "any technological application that uses biological systems, living organisms or derivatives thereof, to make or modify products or processes for specific use" (United Nations Convention on Biological Diversity)..

Hybridization of plants has also been done for millennia. Today's corn, for example, barely resembles the maize of early Native Americans. All of our modern cereals, fruits, and vegetables have their origins in wild grains and produce that our ancestors cultivated to develop the modern varieties that fill our supermarkets' shelves and bins.

Probably the biggest difference between pre-20th century experiments and today's processes is the understanding scientists have today of how genetic manipulation works. Before the discovery of DNA's structure by Watson and Crick in 1953 (see page 27), there was no understanding of how genetic inheritance occurs at the cellular level. Before Watson and Crick, all experiments were simply trial-and-error.

Transgenic Crops

Genetic engineering is revolutionizing agriculture. Plants naturally hybridize in the wild, but such hybridization is completely random. Plant breeders deliberately hybridize plants, but it is a slow process that may take scores of generations to yield results. With genetic engineering, plant breeders can insert individual genes into the genetic material (DNA) of a host plant, resulting in a new strain of that plant in a single generation. The foreign gene that has been inserted causes a different set of proteins to be synthesized. Different proteins can make plants resistant to disease, rust, mold, or mildew; unpalatable to insect pests; resistant to frost and freezing during cold weather spells; able to grow more rapidly; or able to produce more and better quality protein, seeds, and fruit.

Chem Vocab

The term *transgenics* refers to the process in which the genes from one plant are inserted into a plant that is of a completely different species. In transgenics, genes of plants that normally would be unable to hybridize are brought together. *Cisgenics*, also called *intragenesis*, occurs when genes from a plant of the same species are inserted into a host

How Genetic Engineering Works

Scientists' understanding of DNA has progressed to the point that specific pieces of DNA (i.e. specific sequences of genes) can be precisely cut out of the DNA of one species and inserted into a specific location in the DNA of another species. Because DNA has been combined from different species, the resulting material is called recombinant DNA. The cutting of a DNA molecule is accomplished by inserting an enzyme that can recognize specific sequences of genes and severe them from the rest of the DNA molecules.

Important Point

What is incredibly remarkable about genetic engineering is that it is taking place inside the nucleus of a cell, which is so small it can only be seen under a microscope. The genes that make up a DNA molecule are themselves too small to be seen even under a microscope.

Some DNA fragments have "sticky ends" that allow the fragments to link together. "Stickiness" is caused by base pairs at the ends of two chains being complements of each other, i.e. cytosine and guanine, or adenine and thymine. Recall (see chapter 31) that complementary base pairs link together because of hydrogen bonds between the molecules. Unlimited numbers of different kinds of novel genetic material can be formed this way. Just as DNA naturally codes for proteins, the new genetic material codes for specific proteins that increase plant productivity.

> ### Important Point
>
> Genomes
>
> The genetic code stored in DNA is all about protein synthesis. Genetically modifying an organism changes the code so that novel proteins are synthesized.

A common means for inserting new genetic material into a plant's cell nuclei is to use bacteria. A specific bacterium induces the formation of small tumors called crown-gall tumors. Foreign genes can be substituted for the bacterium's tumor-inducing genes. Substituting these foreign genes is the goal of the procedure. For example, in the case of inducing resistance to insect predators, the novel proteins that have been introduced are toxic, or at least unpalatable, to insect larvae normally preying on the plants.

A commonly-used herbicide is the compound glyphosate (trade name Roundup, manufactured by Monsanto). Clinical trials have yielded consistent results that glyphosate is nontoxic to animals, yet extremely effective at killing plants. Unfortunately, glyphosate is not selective about which plants it kills—it kills desirable crop plants just as effectively as it kills weeds. It works by blocking the action of an essential enzyme that all plants must have. To make crops resistant to glyphosate, genetic engineering substitutes a mutant form of the enzyme that is not affected by glyphosate, but which still helps synthesize an essential amino acid required for plants to grow.

The term *genetic engineering* is used along with similar terms that include *genetically modified organisms* ("GMOs") and *recombinant DNA technology*. Whichever term is used, the result is the same—the modification of the sequence of genes in an organism's DNA molecules. Because genes are the template for protein synthesis, changing an organism's genes changes the proteins it manufactures.

The Research Effort

Clearly a tremendous amount of research, which represents a significant financial investment, goes into the development of a successful transgenic plant. First of all, researchers must be convinced that the mutant gene has actually been incorporated into the targeted plant. Then, researchers must be convinced that the mutant gene is in fact producing the desired protein, and that the protein has the effect on plant productivity that was desired in the first place. All this has to be accomplished in a way that economical large-scale production is possible (after all, the company has to make a profit if it is to stay in business).

Chem Vocab

Genetic engineering is the process of inserting genes into, or removing genes from, an organism's DNA molecules to modify that organism. The result is new, presumably desirable, characteristics.

The result of genetic engineering is a *genetically modified organism (GMO)*.

The term *recombinant DNA technology* emphasizes combining DNA from different sources—usually two different species of plants or animals.

Historical Note

The first genetically modified organisms were bacteria, which were produced in 1973. The genetic modification of mice was accomplished in 1974. Genetically modified food has been available in supermarkets since 1994.

Finally—and by no means of least importance—introduction of transgenic plants cannot have harmful effects on the environment, especially on humans. No one wants to buy a box of cereal that is going to make them sick, so extensive trials have to be conducted on both animal and human subjects to rule out adverse side effects before a transgenic crop can go to market.

Feeding 7 Million People

Feeding the world's human population has been a serious issue for hundreds, if not thousands, of years. In 1798, the British economist Thomas Robert Malthus (1766–1834) published *An Essay on the Principle of Population*, in which he predicted that any projected increases in food supply could not possibly keep up with increases in human population. Malthus' thesis was that widespread famines were inevitable.

Large-scale famines did not occur, however, largely due to the increased production of two kinds of agricultural chemicals: fertilizers and pesticides. With them—and now with genetic engineering— food production has more than kept up with population growth, despite the fact that world population has increased seven-fold, from 1 billion people in 1800 to 7.1 billion people at the beginning of 2014.

Fertilizers

Since the dawn of agriculture ten thousand years ago, the most widely used fertilizer has been animal manure, which improves soil texture, holds moisture in soil, and, most importantly, is a readily available source of nitrogen. Planting the same crops on a parcel of land year after year quickly depletes the soil of nutrients. One solution is for some land to be left fallow (i.e. not planted on) for a growing season or two, giving the soil an opportunity to recover. Or, crops can be rotated. Crop rotation takes advantage of the fact that different crops have different nutrient requirements, so by varying the crops that are planted, the same nutrients are not depleted each year. Another solution is to work manure into the soil between growing seasons, although in developed countries, where there isn't enough manure to go around, adding chemical fertilizers is more common.

Nitrogen is the most important nutrient required for plant growth, and many fertilizers are sources of nitrogen in various forms. With the development of the Haber process for producing ammonia (see page 278), ammonia-based fertilizers became readily available on a large scale. Today, the most common ammonia-based fertilizers are liquid ammonia, which can be injected directly into soil at the depth at which roots are growing, and ammonium nitrate (NH_4NO_3), which is packaged in pellet or granular form and is readily available at your local garden center.

Other important plant nutrients found in fertilizers include phosphorus, potassium, calcium, magnesium, and sulfur. Different fertilizers are applied to different crops and at different times of year. All these fertilizers cost money and are a significant source of pollution.

In traditional agricultural operations the purchase and application of fertilizers are a major expense. Contributing to the high cost is the problem that the process of applying fertilizer to soils is not very efficient. Significant amounts of fertilizers accompany surface run-off of water and are washed from the fields into adjacent aquatic ecosystems. Nutrient overload creates a myriad of environmental problems such as algal blooms and fish kills.

Important Point

Adding nitrogen and phosphorus to lakes and ponds causes algae and the population of microorganisms that feed on algae to skyrocket. The consumption of oxygen by these organisms is measured by biological oxygen demand (BOD). High levels of oxygen consumption decrease the oxygen that is available to higher organisms such as fish and mollusks which may die without sufficient oxygen.

Pesticides

Crop yields can also be increased by using pesticides, although pesticides are expensive and create their own problems. In agricultural applications pesticides come in two major groups: insecticides, which kill insects, and herbicides, which kill weeds. Much of modern agriculture is of the kind referred to as *monoculture*, which is the production of a single crop over hundreds or thousands of acres. Monoculture, however, sets the dinner table for a large host of insect pests. Without the application of insecticides, the loss of food and fiber crops to insects would run in the billions of dollars a year. The same

Chem Vocab

Fiber crops are crops that typically are not consumed as food. An important fiber crop is cotton, which is mostly used for clothing.

fertilizers that promote the growth of desirable crops also stimulates the growth of weeds, which in turn compete—maybe even outcompete—with crops for nutrients. It is desirable, therefore, to develop selective herbicides that kill weeds without harming the desired crops.

Important Point

Monoculture

If you live in or have traveled through the Midwest, you have witnessed monoculture. When driving highways through Nebraska, Iowa, South Dakota, or Illinois, for example, you see hundreds of thousands of acres of farmland devoted exclusively to growing corn.

The problem with pesticides is that they are poisons. As such, they often are as toxic to humans—small children especially—and pets as they are to insects and weeds. Nowhere has this point been emphasized with greater impact than by the biologist Rachel Carson (1907–1964). Carson's environmental classic *Silent Spring*, published in 1962, alerted the world to the widespread ecological harm being done by the insecticide DDT.

Historical Note

Silent Spring was an extremely controversial book. Rachel Carson was vigorously attacked both personally and professionally by the pesticide industry with claims of incompetency and false accusations. In 1972, however, the U.S. Environmental Protection Agency (EPA) banned the use of DDT in the United States except in cases of genuine emergency.

One Solution: Genetic Engineering

An alternative to applying more and more fertilizer and more and more pesticides is to genetically modify plants. The world has witnessed tremendous growth since the 1970's in technologies that alter the DNA of plants. One approach is to make plants better able to fix nitrogen and to produce more seeds and fruits, making them less dependent on fertilizers. Another approach is to make them produce their own insecticides, making plants more resistant to the ravages of insects. Food plants can also be made resistant to the herbicides used to kill weeds.

Meeting the World's Demand for More Food

Most of Earth's surface area is covered by water. Significant portions of Earth's surface are covered by deserts, or are too high in elevation for agriculture, or the terrain is too rugged. Consequently, most arable land is already under cultivation. Much of the success of agriculture can be attributed to irrigation, but the majority of irrigation systems that can be built have already been built. More fertilizer or more pesticides is economically unfeasible in developing countries and both are acknowledged as significant sources of pollution in developed countries. In practical terms, the only real way to achieve further increases in the world's food supply is to improve the production of existing crops and animals.

In addition to increasing the supply of food itself, it is also essential to improve the quality and quantity of protein in food. Plants that are rich in protein tend to have high nitrogen requirements, which traditionally have been met by adding more fertilizer. Historically, selective breeding has produced hybrid strains of important crops like corn and wheat. Today, however, the strategy to solving the problem of growing more and better quality food is to develop new strains of food crops through genetic engineering.

In recent years new strains of plants have been developed, including cassava (an oil-producing plant), tomatoes, soybeans, rice, corn, potatoes, papaya, sugar beets, squash, alfalfa, and sativa (which produces oils similar to fish oil), as well as non-food crops like tobacco, roses, and carnations.

Vegetable Oils

Vegetable oils flavor our food and are used in cooking. In addition to cassava, other oils that are produced by genetic engineering include cottonseed oil and canola. Even if you do not add vegetable oils to your food, chances are good that oils from GMO's are in many of the foods you eat, e.g. potato chips, margarine, and packaged foods.

Besides growing cotton for clothing, cottonseed oil comes from cotton. Ninety percent of cotton grown in the United States has been genetically modified.

Tomatoes

A major reason for genetically modifying tomatoes is to make them last longer. Unless tomatoes are labeled as "vine-ripened" or "hot-house grown," the tomatoes sold in most supermarkets are picked while still green and long before they have had a chance to ripen. Picking tomatoes prematurely allows time for the tomatoes to be transported to markets without rotting. Genetically modified tomatoes, on the other hand, rot much

Important Point

Amino Acids

Plants that are rich in protein require nitrogen because nitrogen is an essential component of amino acids. Proteins consist of long chains of amino acids.

more slowly. They can remain on the vine until they have ripened, and they last better during transportation to markets.

Soybeans

It may surprise you to learn that over 80–90 percent of soybeans grown in the United States are genetically modified organisms. Considering how much soy is consumed by vegetarians, for whom soy products are substitutes for meat, that is a lot of GMOs. Even people who are mostly meat-eaters probably consume a lot of genetically modified soy products since most soybeans go into feed for farm animals. Also, even if you do not intentionally choose soy products, soy is found in a wide variety of other common foods, including ice cream, cereal, and baked goods. If you want soy in your diet—but not the genetically modified kind—you can consume tofu and soy sauce and a few other non GMO soy products, which at the present time, at least, are made from non-genetically modified soybeans.

Rice

It is estimated that at least half of the world's population subsists entirely, or almost entirely, on a diet of rice. Among the world's poorest people, a day's food supply is, at most, a bowl of rice. Unfortunately, rice lacks several important vitamins and minerals. For example, rice lacks vitamin A, and vitamin A deficiency is a leading cause of blindness. Rice is also low in, or lacking entirely, the mineral iron. Iron deficiency causes anemia and contributes to babies born with low birth weights. The reason for developing genetically modified varieties of rice is to produce rice that contains beta-carotene (which the body subsequently converts to vitamin A) or iron. Varieties of rice have been produced that contain one or the other (beta-carotene or iron), but not both. Research is continuing in that area.

Corn

Most corn grown in the United States goes into animal feed. Some goes into producing biofuels—ethanol especially—to supplement gasoline supplies. The rest is consumed by humans directly. The main reason for genetically modifying corn is to make corn resistant to insect predation. By making corn produce its own insecticides, the need to spray fields of corn with insecticides is reduced substantially. Half of all corn grown in South Dakota has been genetically modified, with significant quantities of genetically modified corn also being grown throughout the rest of the United States. Whether you eat corn directly, eat meat from farm animals that have been fed with corn, or just drink soft drinks that contain high fructose corn syrup, you are undoubtedly ingesting genetically modified corn.

Potatoes

Potatoes originally were cultivated in the New World by improving upon wild varieties. From the Americas, potatoes have spread throughout the world, especially into Europe and Africa. (Potatoes are less important in Asian countries, where rice is still the dominant food crop.) People only eat about one-quarter of the potatoes that are grown worldwide; most potatoes go into animal feed and into the manufacture of starch. Potatoes are very susceptible to disease, as was demonstrated by the Irish potato famine in the 1800s, so research is being done to develop disease-resistant potatoes. Also, because potatoes are eaten by many of the poorest people in the world, research is being conducted on developing genes that can be injected into potatoes and that would act as vaccines against disease.

Papaya

Farmers in Hawaii came close to losing all their papaya crops to disease. Papaya was only saved by the introduction of genetically engineered variety. In May 2013, however, a bill was introduced on the island of Hawaii to ban all GMOs. Pointing to the success of genetically engineered papaya, subsequent bills have been introduced that would exempt genetically engineered papaya from the ban on GMOs.

Sugar Beets

Natural sweeteners are derived from corn syrup, sugar beets, and cane sugar. Almost all corn and sugar beets have been genetically modified. Cane sugar does not contain GMOs.

Other Crops

It is probably accurate to say that virtually any plant that has commercial importance is under investigation to produce genetically modified varieties that can be grown in greater quantity and more economically. Tobacco-growing, for example, is a big industry with huge financial rewards to farmers who can grow the most crops. Developing insect-resistant strains of tobacco reduces destruction of crops by insect pests.

GMOs in Animals

In the case of farm animals, it is unnecessary to genetically modify the animals themselves to make them grow more rapidly. What is more common is to give farm animals feed that has been genetically modified—corn, soybeans, and alfalfa are widespread examples.

Not all genetic engineering is done with terrestrial organisms. Growing aquatic animals is called *aquaculture*. In the wild, salmon require several years to reach full maturity. Research is underway to produce genetically modified salmon that mature much more rapidly and are larger at maturity than non-genetically modified salmon. In cultures where marine species of fish are important sources of protein, the production of genetically modified salmon on fish farms could make an important contribution to people's diets. In addition, producing salmon on fish farms could protect wild populations from overfishing.

To GMO or Not to GMO

Recall that this chapter began with the argument that a growing world population needs both more food and better quality food. An important goal of producing transgenic food crops is to meet that demand. Another important goal is to meet that demand while at the same time reducing the need for fertilizers and pesticides. The potential benefits of transgenic crops are enormous in a world with 7.1 billion humans—a number that could double by the end of the twenty-first century.

Unfortunately, genetically modified organisms are not without risks. You have probably seen bumper stickers that proclaim "No GMOs." Some municipalities have banned GMO's from their grocery stores. Garden stores advertise that there seeds have not been genetically modified. An increasing number of companies have jumped on the no-GMO band wagon by labeling their products "GMO-free." Chickens are advertised as having been "free range" rather than grain-fed, presumably because natural vegetation should be free of GMOs. There is no benefit to eating GMOs, so it is understandable to question their presence in food.

In January, 2014, General Mills Inc. announced that its original-flavor Cheerios would now be manufactured without any genetically modified organisms. Since there are no genetically modified strains of oats, it is just necessary not to add any cornstarch, sugar, or other ingredients that have been genetically modified. General Mills' rationale for making the change is that original Cheerios are often a baby's first finger food, so it should be as pure as possible.

There are national organizations that are working to ban all GMOs in food. Right now companies have the option of labeling their food "GMO-free" if that is the case, but so far in the United States there is no law requiring food that contains GMOs to be labeled as containing GMOs. Manufacturers argue that having to label food as containing GMOs would attach a stigma to GMOs that they feel is unwarranted.

Besides uncertainty about the effect of GMOs on human health, of considerable concern is the fear that transgenic organisms will hybridize with wild organisms, decreasing biological diversity. Nothing can be done to prevent natural plants from being contaminated by plants in nearby fields. In the case of plants that are made resistant to herbicides, for example, they could cross-fertilize with plants growing in nearby fields and render them resistant to herbicides also. Potentially, the very weeds farmers are trying to control could themselves become immune to herbicides.

The immediate concern to most consumers, though, is whether or not transgenic plants and animals are healthy to eat. Nobody knows what the repercussions of this technology are going to be. There is insufficient data to know yet whether or not GMOs are safe for human consumption. GMO opponents have only to point to DDT, asbestos (used in building insulation), thalidomide (taken by pregnant women), and "fen-phen" (a weight-loss drug) as examples of seemingly benign substances that turned out to have severe adverse health effects. DDT exacted a heavy death toll on animals at all levels of the food chain and on human health. Asbestos workers often succumbed to lung disease. Thalidomide resulted in severe birth defects. Fen-phen was linked with damage to heart valves.

It is not unreasonable to ask whether genetically modified plant and animal foodstuffs will take their toll on human health after decades of consumption. Of great concern are possibly adverse health effects on developing fetuses, infants, and persons who are already weakened by other health issues.

Conclusion

Should the production of GMOs be expanded or outlawed? That is a really tough question to answer. If persons are well-fed and in fundamentally good health, it is easy to say GMOs should be banned. But if parents are suffering the agony of watching their children die from the effects of malnutrition, it is difficult to keep food from them that could save their children's lives. Clearly, there is no easy answer. If you are concerned, shop for products that are GMO-free.

CHAPTER 33

What's In Your Medicine Cabinet?

In This Chapter

> Infectious diease

> Vaccines

> Synthetic drugs

> Genetically engineered drugs

For thousands of years humans experienced low life expectancies (a few decades at most), high rates of infant mortality, high rates of mortality among women during child birth, and low rates of children surviving to adulthood. It was well into the 19th century before the germ theory of disease began to be understood. Babies were delivered, sick patients treated, and limbs amputated—all without the benefit of sterile or even moderately sanitary conditions.

Infectious diseases were rampant, whether disease was transmitted by insect bites or other vectors or resulted from infections due to bodily injuries that could not be treated. There were no antibiotics and no vaccines. Today's eighty-year life expectancies are due in part to better nutrition, but they are also largely due to an understanding of germ theory and to the incredible advances that have occurred in in medicine to treat and cure disease.

For millenia the human race has obtained pharmaceuticals from natural sources. While naturally-derived drugs have done remarkably well in advancing the overall state of human health, there are many diseases for which no sources of natural remedies are known. In this chapter you will learn about the progress that has been made in recent decades in the development of pharmaceuticals using genetic engineering.

Chem Vocab

A pathogenic vector is an organism that transmits disease from one organism to another organism.

Genetically Engineered Food versus Genetically Engineered Drugs

There is a distinct difference between the issues of genetically modified organisms in food and genetically engineered drugs used to combat disease. GMOs in food are consumed by thousands, if not millions, of people who probably are completely unaware that GMOs are in their food. Opponents of GMOs argue that the consumption of GMOs is a huge, uncontrolled experiment. No one is actively monitoring the health of consumers or checking them periodically to see what side effects, if any, may be developing.

In contrast, genetically engineered drugs are administered to specific individuals who have been diagnosed as suffering from specific diseases. These drugs must be prescribed by a patient's medical doctor. Ideally, doctors should be fully aware of the source of any drug they prescribe, the drug's potential side effects, and what alternatives may exist. Doctors should also be monitoring their patients' progress and being on the lookout for adverse side effects that appear in their patients. Because of the relative newness of genetically engineered drugs, any application of these drugs is still experimental, but not on the large scale of genetically modified foods.

Unlike the consumption of GMOs, uses of genetically engineering drugs are relatively controlled experiments. People consuming GMOs are unknowing or involuntary participants in the "experiment." With genetically engineered drugs participation in the "experiment" is voluntary on the part of the patient, even if it seems that no alternative treatment exists. Patients should always be partners with their doctors in determining the best courses of action to follow for optimum health. If doctors explain to their patients the nature of any drug they prescribe and what the potential consequences are of taking the drug, patients can choose whether or not to take that drug or to request a different treatment. Hopefully, after reading this chapter, you will be in a better position to discuss with your doctor any drugs you might be taking and, if applicable, to decide if a genetically engineered drug is the right choice for you.

Infectious Disease

In the previous chapter it was emphasized that abundances of food have occurred primarily in the richest countries in the world. There is not enough access to food in the poorest countries to feed hundreds of millions of people, often because there simply is not an efficient distribution system. Consequently, deaths due directly or indirectly to hunger or malnutrition are still common. In the poorest countries people may also have no access to doctors or antibiotics or vaccines. The drugs they need exist, but there is no money to pay for them, or no system exists to distribute drugs to remote villages, or there are no doctors to administer them.

Three infectious diseases that often have been described as the "fourth horseman of the apocalypse" (a reference to the book of *Revelation*) are AIDS, tuberculosis, and malaria. AIDS ("acquired immune deficiency syndrome") is a relatively recent scourge, but people have suffered from tuberculosis (TB) and malaria for all of human history. Just these three diseases alone result worldwide in an estimated half a billion new cases and at least 5 million deaths every year. In a world of 7.1 billion people, half a billion cases of disease means that one out of every 14 people in the world is infected with one of these three diseases *every year*. If the half billion cases were distributed evenly over the entire world, one of every 14 people you know (and possibly you yourself) would be infected. The vast majority of cases, however, occur in sub-Sahara Africa and other poor regions of the world—relatively out of sight of affluent North Americans.

The current estimate is that one-third of the world's population is infected with the bacterium that causes tuberculosis—that's over 2 billion people. In developed countries like the United States, infected individuals can be administered lifesaving antibiotic treatments. Once infected, a person will still test positive for tuberculosis for the remainder of his or her life, but the bacterium will remain dormant. In poor countries, where people often have no access to antibiotics, their health often continues to worsen until they die.

Worldwide, malaria is responsible for 8 percent of deaths of children under the age of five. Each year 1.5 to 2.7 million people die from malaria. Because malaria is transmitted by mosquito bites, malaria is prevalent mostly in warm tropical or subtropical regions of the globe that have high mosquito populations. Because malaria has largely been eradicated in developed countries, virtually all of these deaths occur in the world's poorest countries. Presently, there is no vaccine with which to completely inoculate people against malaria, although there are some drugs—including antibiotics—that can lessen the risk of contracting malaria.

Chem Vocab

Bacterium is the singular form of *bacteria*. Bacteria are one-celled organisms, many of which are beneficial or benign to humans, while many are pathogens.

Viruses are simpler organisms. Consisting of a single nucleic acid surrounded by a protein molecule, a virus has to reside in a living host in order to function biologically.

HIV (the human immunodeficiency virus that is responsible for AIDS) currently infects 33 million people, of whom 70 percent live in Africa. Since 1981, more than 25 million people worldwide have died of AIDS. In countries of sub-Saharan Africa, the infected rate can be anywhere from 10 to 30 percent of the population. And Africa is not alone—India is number two for AIDS cases. Rates of infection in Latin America and the Caribbean are beginning to catch up with Africa. There are an estimated 1 million cases of AIDS in the United States (out of a population of about 317 million people). Worldwide there are 2 million deaths from AIDS every year. In Africa, an estimated 15 million children have been orphaned because both parents have died of AIDS. Many of these children themselves suffer from AIDS.

These are not the only diseases. Developing countries have high rates of malnutrition, lack of sanitation, exposure to parasites, and contaminated water that all contribute to disease. These conditions are debilitating enough by themselves to cause death, or at least to be a contributing factor. Contaminated water alone kills an estimated 5 million people a year. Both adults and children who suffer from these conditions are so weakened that they do not have much chance of survival if they also become infected with AIDS, tuberculosis, or malaria.

In developed countries many infectious diseases have been conquered, but there is still cancer, heart disease, diabetes, Parkinson's disease, Alzheimer's disease, and insect-borne diseases like Lyme's disease and West Nile virus. Progress has been made after decades of cancer research, but cancer is by no means conquered. Prevention or cures for other life-threatening diseases still lie in the future.

Early Vaccines

The purpose of a vaccine is to make a person immune to specific viruses or bacteria that cause disease. These viruses or bacteria collect-ively are called *antigens* or *pathogens*. How the body fights antigens is a complex subject. However, a very common response to the presence of antigens is to produce protein molecules called *antibodies* that can neutralize or destroy the antigens. Poisons such as snake venom or insect toxins can elicit similar responses. Antibodies are commonly produced by the body's white blood cells.

Viruses and bacteria are constantly mutating. Influenza (the "flu") is caused by several different kinds of viruses that are mutating all the time. Therefore, flu vaccines also have to change. Last year's vaccine may have been effective against last year's virus, but will have no effect on this year's virus.

Chem Vocab

To be *immune* to something means to be unaffected by it or unresponsive to it. A person who is immune to an illness will not acquire that illness.

Smallpox Vaccine

The first vaccine was developed in 1796 by the British physician Edward Jenner (1749–1823) as a preventative procedure against smallpox. Smallpox had been a scourge for all of human history. Epidemics were frequent and widespread. It was recognized, however, that people who contracted smallpox and who did *not* die would never contract it again. In other words, once a person's body had successfully defeated smallpox, that person was immune to it for the rest of his or her life.

Important Point

Antibodies

Most antibodies are specific to only one kind of antigen. The body has no defense against a particular antigen if the body cannot produce the specific antibody required to fight that antigen.

Jenner's insight was to inject people with small amounts of cowpox, which is milder and less debilitating that smallpox. The bodies of people who were injected with cowpox produced antibodies that were also effective against smallpox, and therefore would never get smallpox. Once the antibodies had been produced, they would remain in a person's body for the rest of that person's life. Smallpox vaccinations became routine. In 1980, the World Health Organization declared that smallpox had been completely eradicated from the globe. Because smallpox is no longer a threat, smallpox vaccinations are no longer given. The eradication of smallpox is considered to be the first successful case of completely eliminating a disease that had been rampant worldwide.

Rabies Vaccine

Another early vaccine was developed by the French chemist Louis Pasteur (1822–1895). Pasteur reasoned that if pathogenic microbes were sufficiently weakened, by heating them for example, and then injected into a patient, the patient's own immunological system would be able to produce antibodies that were strong enough to kill the microbes. Once those antibodies had been produced, they would continue to produce immunity (resistance) to that disease, often for the rest of the patient's life. Pasteur first experimented with vaccinating sheep against anthrax (a particularly virulent, infectious, and usually fatal, disease). His first successful human inoculation was for rabies and prevented a young boy from getting rabies after he had been bitten by a rabid dog.

Historical Note

Louis Pasteur is also famous for the development of the process that uses controlled heat to kill harmful microbes in wine, milk, and beer. In Pasteur's honor this process is called *pasteurization*. Pasteur died only a few years before the implementation of Nobel Prizes. If he had lived long enough, it is reasonable to assume that he would have been one of the first recipients of the Nobel Prize in chemistry.

Polio Vaccine

A third example of a crippling disease that was conquered by a vaccine is polio (more accurately referred to as *poliomyelitis*). Polio is an infectious disease that causes paralysis. Paralysis of the lungs can lead to death. In 1955, the American research scientist Jonas

Salk (1914–1995) developed a vaccine for polio that contained dead viruses. (Even though the viruses are dead, the body still produces antibodies to fight them.) A few years later, the American medical researcher Albert Sabin (1906–1993) developed an oral polio vaccine. The widespread use of the two vaccines has virtually eliminated polio in developed countries.

Historical Note

Franklin Roosevelt (1882–1945), the 32nd President of the United States, became crippled with polio at the age of 39. Despite the fact that he was never able to walk unaided again, Roosevelt was elected President four times and guided the United States through World War II. He died of a cerebral hemorrhage shortly into his fourth term of office.

Other Common Vaccines

In the United States and Canada, children today are given vaccines to prevent a host of diseases, including diphtheria, whooping cough (pertussis), tetanus, measles, mumps, chicken pox, meningitis, and polio. People are so unfamiliar with some of these diseases that most people do not even know what they are anymore. Adults of all ages are often vaccinated against influenza, hepatitis A and hepatitis B. Older adults are advised to be vaccinated against shingles and pneumonia. Vaccination programs have been phenomenally successful at contributing to people's good health.

Unfortunately, these vaccinations may not be available to millions of people in developing countries. Also, there are many diseases, such as malaria for example, for which there are no vaccines. In addition to vaccines, some drugs may be too expensive for people to purchase them, even people with health insurance. Most leukemia drugs, for example, can cost a patient $100,000 a year for treatment, far beyond the means for most people to buy them.

Genetically Engineered Drugs

Most of the drugs that have been described come from natural sources—plants or animals. However, there may not be sufficient supplies of the plants or animals to satisfy demand. Drugs may have to be synthesized, hence the interest in genetically engineered drugs. Using biotechnology it is often possible to produce supplies of a drug that are sufficient to meet demand. Although it is unlikely that anyone expects to ever develop a "super-drug" that would combat all diseases, developing a highly effective drug for each disease is surely a worthwhile goal. The following examples of genetically engineered drugs give an indication of successes that have been obtained in the never-ending battle against debilitating and life-threatening diseases.

Insulin

Diabetes results from a pancreas that does not function properly. A normal pancreas produces hormones that aid in metabolism, which is the breakdown of the food we eat. The most familiar hormone produced by the pancreas is insulin, which regulates the concentration of the simple sugar glucose in the blood stream. When the pancreas is not producing sufficient insulin, then too much glucose accumulates in the blood stream; the result is the condition called diabetes. In severe cases poor circulation can result in blindness, gangrene in the lower extremities (feet and legs) that may necessitate amputation, heart disease, stroke, or death.

Important Point

Insulin

Insulin molecules are proteins. Since DNA is the template for synthesizing proteins, altering DNA in organisms to produce insulin identical, or at least very close to, human insulin is an example of the basic principle that the goal of genetic engineering is the production of protein molecules that are beneficial to human health.

Diabetics (individuals who suffer from diabetes) may require injections of insulin to help regulate blood sugar levels. The earliest sources of insulin were the pancreas of cattle and pigs. Some insulin preparations may also be prepared from a combination of beef and pork sources. Pork insulin very closely resembles human insulin. Insulin from pork, however, is unacceptable to followers of certain religious faiths: Jews and Muslims, for example, eschew all pork products. Also, some people suffer allergic reactions when given beef-or pork-based insulin.

More recently, synthetic insulin has been one of the successes of recombinant DNA technology. Examples are marketed under the trade-names Humulin and Novolin, which essentially mean "human insulin" and "new insulin."

Antibiotics

One of the first successful antibiotics to be discovered was penicillin, which is obtained from the Penicillium mold. The Scottish bacteriologist Sir Alexander Fleming (1881–1955) discovered by accident that the bacterium Staphyloccus was killed by Penicillium mold. Penicillin could not be produced from mold in large quantities, however. Over the next two decades, processes for producing penicillin on a large scale were developed. Shortly after penicillin began to be produced in large quantities, however, it was discovered that microbes quickly acquired resistance to it.

In 1955, the American researcher Lloyd Conover (b. 1923) introduced tetracycline, which soon became the most widely prescribed antibiotic in the United States. In particular, tetracycline is a common substitute for individuals who are allergic to penicillin.

Unfortunately, microbes' resistance to antibiotics is on the increase. Despite the fact that billions of pounds of antibiotics are used around the world every year, their effectiveness against some microorganisms is decreasing. One reason may be that antibiotics are now widely added to animal feed to produce healthier animals. Residual amounts of these antibiotics are still in meat when the meat is consumed.

Another reason is that antibiotics are often prescribed routinely, even when they may not be effective. People often go to the doctor expecting to walk out of the doctor's office with a prescription, even if it is for a common cold, which is caused by a virus and for which an antibiotic has no effect. Also, people who have a legitimate need for an antibiotic often stop taking their antibiotic as soon as they

Important Point

Humulin is genetically engineered insulin that is identical to human insulin. An important characteristic of Humulin is that persons who suffer allergic reactions to insulin derived from beef or pork are not allergic to Humulin.

Important Point

Staphyloccus and streptococcus are two common bacterial infections occurring in humans. Usually, the abbreviations *staph infection* or *strep infection* (as in "strep throat") are used start to feel better. However, not all of the bacteria causing their illness have been killed yet. Bacteria that have not been killed have at least some resistance to the antibiotic. As those bacteria begin to multiply again, the next generation consists of bacteria that are antibiotic-resistant.

The more antibiotics bacteria in our bodies are exposed to, the more likely the bacteria will become resistant to the antibiotics. As a result, of alarming concern in the medical community is the evolution of "superbugs" that are resistant to an increasing wider spectrum of antibiotics.

Important Point

Superbugs

Superbugs have become so resistant to any antibiotics on the market that they are not killed by any of them. According to the Centers for Disease Control and Prevention (CDC), superbugs could become one of the most urgent healthcare issues of the current century.

Enter genetically engineered antibiotics—drugs to which superbugs are not resistant. One of the success stories has been a novel form of erythromycin, which is also a substitute for penicillin. Today, antibiotics produced by recombinant DNA technology represent a multi-billion dollar industry.

Anticoagulants

Blood clotting, or coagulation, is a natural response to cuts and lacerations of the skin and prevents a victim from bleeding to death. In some situations, however, blood clotting is undesirable because it reduces or prevents the natural flow of blood through arteries and veins. Anticoagulants reduce blood clotting. Doctors administer anticoagulants to prevent blood clots from forming, helping to prevent blood clots in the lungs, heart attacks, and strokes.

Important Point

Hemophilia

The opposite problem of blood clotting when it should not is blood not clotting at all. In situations of hemophilia (literally meaning "blood-loving"), blood does not clot at all. Hemophiliacs have to be careful to avoid any cuts or bruises since bleeding does not stop once it has started. There are drugs to treat hemophilia, too.

An anticoagulant sold under the brand name ATryn is manufactured by the American biotechnology company GTC Biotherapeutics. The company has genetically modified goats that produce milk con

taining proteins with human anticoagulant properties. Human trials have been successful, and there is no evidence of any harmful effects to the goats.

Interleukins

The principal function of white blood cells is to combat infections. In fact, if a blood test indicates high levels of white blood cells in a patient, that is a sign the patient's body is fighting an infection. Interleukins are substances that can be extracted from white blood cells and that have the same ability to fight infection. Interleukins are in use to combat some forms of cancer. Clinical studies with genetically engineered interleukins have shown some success against cancer.

Interferons

Natural interferons are proteins produced by cells to fight viral infections. Recent research suggests that some interferons may be effective agents in the treatment of diseases such as cancer and multiple sclerosis. The biggest problem with interferon therapy is producing supplies of interferons that are sufficient to meet demand. Clinical trials are ongoing.

HIV/AIDS

A potential HIV vaccine has been developed in Canada under the leadership of Dr. Chil-Yong Kang. Phase I clinical studies completed in August, 2013, used a genetically modified human immunodeficiency virus indicated higher level of antibodies in HIV-free subjects without any serious side effects. Phase II clinical studies were set to begin in 2014 in Canada, the United States, and Europe to extend the studies to persons at high-risk of developing HIV either due to occupational hazards or social lifestyles.

Malaria

At the Walter and Eliza Hall Institute of Medical Research in Australia, a genetically-modified form of the malaria parasite has been created. The modification consists of removing the two genes in the parasite that are responsible for infecting the liver in humans and causing the disease to spread throughout the blood stream. At the time of publication, a vaccine made from this modified parasite is undergoing clinical trials at the Walter Reed Army Institute of Research in Bethesda, Maryland. Primary financial support for the trials has been provided by the Bill and Melinda Gates Foundation.

Lyme Disease

Lyme disease is transmitted by deer ticks. Symptoms include fever, headaches, fatigue, muscle and joint pain, and—in severe cases—heart palpitations and neurological issues. Scientists at the U.S. Center for Disease Control and Prevention at Fort Collins, Colorado, are developing a genetically modified vaccine that knocks out the gene responsible for the transmission of the bacteria that cause Lyme disease from ticks to humans.

Tuberculosis

An increase in tuberculosis infections has accompanied the emergence of the HIV/AIDS epidemic. Strains of tuberculosis-carrying bacteria have developed resistance to traditional drugs used to treat or vaccinate against tuberculosis. Studies are underway to genetically alter the DNA of tuberculosis-carrying bacteria to produce more effective antibiotics.

Hepatitis B

The hepatitis B virus is a major cause of liver disease around the world with about 200,000 new cases occurring each year in the United States. Conventional hepatitis B vaccines were produced from human blood. Contaminated blood carries a significant risk of infection from other blood-borne diseases like AIDS. Genetically engineered hepatitis B vaccines have been produced in large quantities and are now used routinely. A vaccination regimen consists of three separate injections spaced several months apart.

Conclusion

Genetically engineered drugs have the potential of saving hundreds of thousands, if not millions, of lives a year. Probably every disease for which a vaccine can potentially be developed is under investigation. The goal of reducing human suffering is a laudable one. Given the successes biotechnology has had so far, it is to be expected that growth in the field of genetically engineered drugs will continue to increase.

INDEX

ABOUT THE AUTHOR

Brian Nordstrom quickly developed an interest in the physical sciences due to his excellent physics, chemistry, and astronomy professors at the University of California, Berkeley, during the 1960s. The more he studied those subjects, the more he wanted to learn. Eventually he graduated with an interdepartmental major in physical science. It was difficult for him to choose a single field for graduate work, but he decided on physical chemistry as the best blend of his interests in chemistry and physics.

Returning to Berkeley in the early 1970s, Brian joined the research group in atmospheric chemistry led by Professor Harold S. Johnston. Both undergraduate and graduate studies introduced him to many of the key scientists who were responsible for major discoveries in the sciences during the second half of the twentieth century. His scientific interests evolved to encompass chemical kinetics, environmental science, computational chemistry, and the history and philosophy of science.

After receiving his Master's degree in chemistry in 1975, Brian joined the chemistry faculty at California State University, Chico. The teaching bug "bit" him, and he's been a college chemistry professor ever since. He joined the faculty at Embry-Riddle Aeronautical University in Prescott, Arizona, in 1980, where he has risen through the ranks to full professor and served for several years as the associate dean of the College of Arts and Sciences.

In 1989, Brian completed a doctorate in education at Northern Arizona University, Flagstaff, with an emphasis on college curriculum and instruction. His research in chemical education, chemical kinetics, and computational chemistry has resulted in several publications and conference presentations.

When not pursuing his professional interests, Brian enjoys traveling, camping, hiking, birding, nature study, photography, and reading. His travels these days include visits with his four children, their spouses, and his eight grandchildren.